Objective PLANT SCIENCE At a Glance

The Authors

Deepak Kumar Verma
*M.Sc. (Agricultural Biochemistry), **Ph.D. (Pursuing),
*Department of Agricultural Biochemistry, Collage of Agriculture,
Chandra Shekhar Azad University of Agriculture and Technology,
Kanpur (Uttar Pradesh) – 208 002, INDIA
**Food Process Engineering, Department of Agricultural
and Food Engineering, Indian Institute of Technology (IIT),
Kharagpur (West Bengal) – 721 302, INDIA

Dr. Bavita Asthir
Sr. Biochemist *cum* Head, Department of Biochemistry,
Collage of Basic Science and Humanities
Punjab Agricultural University, Ludhina – 141 004, INDIA

Dr. Dinesh Narayan Bharadwaj
Professor, Department of Genetics and Plant Breeding,
College of Agriculture
C. S. Azad University of Agriculture
and Technology, Kanpur – 208 002, INDIA

Dr. Santanu Kumar Swain
Professor of Seed Science and Technology
Department of Seed Science and Technology, College of Agriculture,
Odisha University of Agriculture and Technology,
Bhuvneshavar – 751 003, INDIA

Dr. S. S. Verma
Professor of Seed Science and Technology,
Department of Seed Science and Technology,
College of Agriculture,
CCS Haryana Agricultural University,
Hisar – 125 004, INDIA

Objective PLANT SCIENCE At a Glance

(For ICAR-NET, ICAR-SRF, ICAR-JRF, IFS, IAS, SAU's and Other Competitive Examinations)

Deepak Kumar Verma

Bavita Asthir

Dinesh Narayan Bharadwaj

Santanu Kumar Swain

S. S. Verma

SCIENTIFIC PUBLISHERS (INDIA)

Published by
SCIENTIFIC PUBLISHERS (INDIA)

Jodhpur –
5 A, New Pali Road
P.O. Box 91
Jodhpur - 342 001 INDIA

Delhi –
4806/24, Ansari Road
Daryaganj
New Delhi - 110 002 INDIA

E-mail: info@scientificpub.com
Website: www.scientificpub.com

Print: 2018

ISBN: 978-81-7233-888-6

Printed in India

Indian Institute of Pulses Research (IIPR), Kanpur - 208 024 (INDIA)

Dr. N. Nadarajan (Director)
Phone : 91-512-2570264 (O), 91-512-2570111 (R)
Fax : 91-512-2572582
Email : director@iipr.ernet.in, n.nadarajan@gmail.com

Forward

For Indian Council of Agriculture Research (ICAR) and Agriculture Scientist Recruitment Board (ASRB), competitive examinations are the order of the day. All agriculture colleges/ agriculture institutes/ agriculture universities etc. are conducting professional courses to admit students at UG, PG level by common entrance examination. Agriculture Scientist Recruitment Board (ASRB), New Delhi conducts ICAR-NET while ICAR-SRF and ICAR-JRF are being conducted by Indian Council of Agriculture Research (ICAR), New Delhi. All competitive examinations are based on the multiple choice type questions. Multiple choice type questions judge the ability of the candidate to apply his/her most appropriate knowledge acquired during the course study and competitive examinations. The multiple choice questions in the book entitled "Objective *PLANT SCIENCE* At a Glance", is compilation made with great care from several resource text book. This book is very precise and has clear understanding to enable students to face the subject in the examination with confidence. And also for teachers and researchers related to plant science subject to improve the knowledge, to provide a powerful knowledge and feedback for stimulating student's success carrier. This book covers a wide range of topics as given in the syllabus of ICAR-NET, ICAR-SRF, ICAR-JRF, IFS, IAS, SAU's and other competitive examinations.

Besides students, this book is knowledge bank for researchers and teachers related to the plant science subjects to impart the knowledge and feedback stimulating student's successful carrier. The author Mr. Deepak Kumar Verma, Dr. Bavita Asthir, Dr. Dinesh Narayan Bharadwaj, Dr. Santanu Kumar Swain and Dr. S. S. Verma has vast experience as a research scholar,,biochemist, teacher and scientist, respectively. Though, in market, there are several books in Plant Science written by Indian authors, but this book differs from others in that it comprises of questions and answers in all aspects and field of Plant Science. Almost all the basics and fundamental asked questions in ICAR competitive examinations in Plant Science have been covered by the authors. It include subjects like Plant Physiology, Genetics and Plant Breeding, Seed Science and Technology, Plant Biotechnology, Plant Molecular Biology etc. are also covered in a lucid and simplified manner. This book which is a compendium of questions and answers in almost topics of ICAR-NET, ICAR-SRF, ICAR-JRF, IFS, IAS, SAU's competitive examinations in Plant Science. I hope this will certainly prove an adjuvant for ICAR and other competitive examinations.

I strongly recommend this book "Objective *PLANT SCIENCE* At a Glance", to all the students preparing forICAR-NET, ICAR-SRF, ICAR-JRF, IFS, IAS, SAU's and other competitive examinations etc. and congratulate the authors for putting enamors effort for compiling such a useful resource book. I earnestly hope that this book will receive appreciation from students, teachers and researchers.

Date: 26.09.2013
Place: Kanpur

(N. Nadarajan)

PREFACE

Botany is a mother of science which has given birth to the several branches viz. genetics, plant breeding, cytogenetics, cytology, anatomy, morphology, taxonomy, physiology, ecology, plant reproduction, systematic, embryology, plant evolutionary developmental biology etc. each being pursued as a speciality in itself. The study of plant sciences has undergone tremendous changes in the past few decades in all above mentioned areas. Agriculture is a backbone of economy all over the globe. In India, Indian Council of Agriculture Research (ICAR) is playing a pivotal role and Agriculture Scientist Recruitment Board (ASRB) playing a key role in development of agriculture education and research by selecting world class scientist in this field. Therefore the present book 'Objective *PLANT SCIENCE* At a Glance' will be an asset in preparation and qualifying ASRB examination by covering objective questions and answers to prepare good Plant Scientist in all fields of agriculture by covering its wide fields of knowledge.

"Objective *PLANT SCIENCE* At a Glance" is a multiple choice question book specially designed to improve the knowledge of students and to provide them a powerful knowledge and feedback in their progress and future opportunities. Thus, primarily this book can serve as a self-assessment guide for the students who are preparing for competitive examinations specially ICAR-NET, ICAR-SRF, ICAR-JRF, IFS, IAS, SAU's etc. And secondly, meant for those who appearing for UG, PG study in agriculture colleges/ agriculture institutes/ agriculture universities etc. The questions of this book are compiled from standard textbooks on Cytogenetics, Genetics, Plant Breeding, Plant Classification, Botanical Description, Economic Field Crops, Plant Hormones, Gene, Gene Expression and Regulation, Genome and Genomic Organization, Stress and Quality Breeding, Seed Science and Technology, with the syllabus of ICAR-NET, ICAR-SRF, ICAR-JRF, IFS, IAS, SAU's and other competitive examinations. The main objective of this book is to help students to enrich their powerful knowledge of plant science because it is essential for the development of students in the area of agriculture.

We hope that this book certainly will increase the knowledge of agriculture in plant sciences, cracking examinations in various fields/subject (Plant Physiology, Genetics and Plant Breeding, Seed Science and Technology, Plant Biotechnology, Plant Molecular Biology etc.) that are essential now a day to provide complete knowledge for the proper accurate, reliable and quicker assessment. This book will also be very useful for teachers and researchers, related to plant science subject to stimulate the success to the student's carrier.

We request and welcome our esteemed readers of this book to send their valuable suggestions and expectations in further improvement of this book in the next issue.

November, 2013

Deepak Kumar Verma
Bavita Asthir
Dinesh Narayan Bharadwaj
Santanu Kumar Swain
S. S. Verma

This book is dedicated to the memory of remarkable people who inspired and encouraged me long ago:
My Grandparents

Shree Meva Lal Swarankar and

Late Shree Radhika Devi

My Parents

Shree Om Prakash Verma and

Shobhawati Devi

My Teacher

Prof. R. C. Srivastava

Prof. P. S. Kendurkar

Prof. Hari Niwas Mishra

Dr. Mukesh Mohan

Dr. Prem Prakash Srivastav

ICAR-NET Syllabus
Genetics and Plant Breeding

Unit 1: General Genetics and Plant Breeding

Mendelian inheritance. Cell structure and division, Linkage, its detection and estimation. Epistasis. Gene concept, allelism and fine structure of gene. Extra chromosomal inheritance. DNA – structure, function,replication and repair. Genetic code. Gene- enzyme relationship. Replication, Transcription and Translation. Gene regulation in prokaryotes and eukaryotes. Nuclear and cytoplasmic genome organization. Spontaneous and induced mutations and their molecular mechanisms. Crop domestication, evolution of crops and centers of diversity. Emergence of scientific plant breeding. Objectives and accomplishments in plant breeding and the role of National and International institutes. Gametogenesis and fertilization. Modes of sexual and asexual reproduction and its relation to plant breeding methodology. Apomixes,
incompatibility and male sterility systems and their use in plant breeding.

Unit 2: Economics Botany and Plant Breeding Methods

Origin, distribution, classification, description and botany of cereals (wheat, rice, maize, sorghum, pearl millet, minor millets); pulses (pigeon pea, chickpea, black gram, green gram, cowpea, soyabean, pea, lentil, horse gram, lab-lab, rice bean, winged bean, lathyrus, Lima bean; oilseeds (groundnuts, sesamum, castor, rapeseed/mustard, sunflower, Niger, linseed); fibers and sugar crops, fodder and green manures; Breeding methods for self-pollinated, cross-pollinated and clonally propagated crops. Component, recombinational and transgressive breeding. Single seed descent. Populations, their improvement methods and maintenance, Hybrid breeding and genetic basis of heterosis. Ideotype breeding. Mutation breeding.

Unit 3: Genome Organization and Cytogenetics of Crop Plants
Chromosome structure, function and replication. Recombination and crossing over. Karyotype analysis. Banding techniques. In situ hybridization. Special types of chromosomes. Chromosomal interchanges, inversions, duplications and deletions. Polyploids, haploids, aneuploids and their utility. Wide hybridization and chromosomal manipulations for alien gene transfer. Pre and post- fertilization barriers inwide hybridization. Genome organization and cytogenetics of important crop species- wheat, maize, rice, brassica, cotton, Vigna, potato and sugarcane. Principles and procedures of genome analysis. Cytogenetic techniques foe gene location and gene transfer, Construction and use of molecular marker based chromosome maps. Comparative mapping and genome analysis.

Unit 4: Quantitative and Biometrical Genetics
Quantitative characters. Multiple factors inheritance. Genetic control of polygenic characters. Genetic advance and types of selection and correlated response. Hardy Weinberg law. Linkage disequilibrium. Genetic load. Polymorphism. Breeding value, heritability. Response to selection, correlated response. Estimates of variance components and covariance among relatives. Mating designs with random and inbred parents. Estimation of gene effects and combing ability. Effects of linkage and epistasis on estimation of genetic parameters. Maternal effects. Genotype environment interactions and stability of performance. Heterosis and its basis. Mating system and mating design- diallel, line X tester, NC-1, NC-II and NC-III designs, approaches to estimate and exploit component of self and cross pollinated crops. Genotype X environment interaction and stability analysis.

Unit 5: Genetic Engineering and Biotechnological Tools in Plant Breeding
Somatic hybridization, micro-propagation, somachonal variation in vitro mutagenesis. Artificial synthesis of gene. Genetic and molecular markers, generations of molecular markers and their application in genetic analyses and breeding. Molecular markers in genetic diversity analysis and breeding for complex characters. Gene tagging, QTL mapping and marker aided selection. Genome projects and utilization of sequence formation. Vectors.DNA libraries, DNA finger printing, DNA sequencing. Nuclei acid hybridization and immunochemical detection. Chromosome walking, Recombinant DNA

technology, Gene cloning strategies. Genetic transformation and transgenics. Antisense RNA, RNAi and micro RNA techniques in crop improvement.

Unit 6: Plant Breeding for Stress Resistance and Nutritional Quality

Genetic basis and breeding for resistance to diseases and insect-pests. Breeding for vertical and horizontal resistance to diseases. Genetic and physiological basis of abiotic stress tolerance. Breeding for resistance to heat, frost, flood, drought and soil stresses. Important quality parameters in various crops, their genetic basis and breeding for these traits. Role of molecular markers in stress resistance breeding: MAS, MARS and MABB.

Unit 7: Plant Genetic Resources and their Regulatory System; Varietal Release and Seed Production

Plant exploration, germplasm introduction, exchange, conservation, evaluation and utilization of plant genetic resources. Convention on Biological Diversity and International Treaty on Plant Genetic Resources for Food and Agriculture. Intellectual Property Rights. Biodiversity Act. Plant Variety Protection and Farmer's Rights Act. System of variety release and notification. Types of seeds and seed chain. Seed production and certification.

Unit 8: Statistical Methods and Field Plot Techniques

Frequency distribution. Measures of central tendency, probability theory and its applications in genetics. Probability distribution and tests of significance. Correlation, linear, partial and multiple regression. Genetic divergence. Multivariate analysis. Design of experiments- basic principles, completely randomized design, randomized block design and split plot design. Complete and incomplete block designs. Augmented design, Grid and honeycomb design. Hill plots, unreplicated evaluation. Data collection and interpretation.

ICAR-NET Syllabus Seed Science and Technology

Unit 1 : Seed Biology

Floral biology, mode of reproduction, sporogenesis, pollination, fertilization, embryogenesis, fruit and seed development. Apomixis, parthenocarpy, polyembryony and somatic embryoids and synthetic seeds. Seed structure of monocot and dicot. Seed maturation and longevity in orthodox and recalcitrant seed. Chemical composition of seed. Seed dormancy - types, causes and mechanisms of induction and release, factors affecting, methods to overcome dormancy and its significance in agriculture. Seed germination - requirements, imbibition pattern, physiological and biochemical changes, and role of hormones.

Unit 2 : Seed Production

Introduction to crop breeding methods. Variety testing, release and notification. Genetic purity concept and factors responsible for deterioration of varieties. Maintenance breeding. General system of seed multiplication. Seed production agencies. Identification of seed production areas and factors affecting it. Compactarea approach in seed production. Seed production planning, equipment, input and manpower requirement. Factors affecting pollination and seed set viz., temperature, humidity, wind velocity, insect pollinators, and supplementary pollination. Male sterility, self-incompatibility and their role in hybrid seed production. Principles and methods of seed production of varieties and hybrids of cereals like wheat, paddy, sorghum, pearl millet and maize; pulses like chickpea, pigeon pea, green gram, black gram, soybean and cowpea; oilseeds like groundnut, brassica, sesame, sunflower and castor; fiber crops like cotton and jute; vegetables crops like tomato, brinjal, okra, chilli, important cole and cucurbitaceous crops; important forage legumes and grasses and seed crop management, time of harvesting and threshing/extraction methods. Seed production technology of plantation crops like coffee, tea, rubber, cocoa, cardamom and pepper. Disease free clonal propagation of crops like potato, sugarcane sweet potato, tapioca, colocasia, betel vine, fruit crops like mango, citrus, banana, guava, sapota, pineapple, grape, apple, pear, plum, peach, apricot and seed production and clonal propagation of annual and perennial flowers like rose, gladiolus, chrysanthemum, marigold, dahlia, flox and petunia. Clonal standards and degenerations. Micro propagation.

Unit 3 : Seed Processing

Principles of seed processing. Seed drying principles and methods. Pre-cleaning, grading, treatment, pelleting and packaging. Seed invigoration and enhancement treatment and their applications. Seed processing machines like cleaner cum grader, specific gravity separator, indented cylinder, seed treater, weighing and bagging machines, their operation and maintenance. Seed quality maintenance during processing.

Unit 4 : Seed Quality Control

Seed legislation - Seeds Act 1966, Seed Rules 1969 and New Seed Bill 2004, Seed Law Enforcement. Seed certification – history, concept, organization, phases and minimum certification standards. Field inspection principles and methods. Inspection at harvesting, threshing and processing. Pre-and post quality testing or genetic purity. Seed Certification Schemes, concepts and procedures. Seed Testing concepts and objectives, its role in seed quality control. Seed sampling, seed moisture testing, purity analysis, germination testing, tolerance tests and equipment. Seed testing procedures for principal agri horticultural crops. Quick viability tests. Seed vigour, its significance and testing methods. Testing for genuineness of varieties – principles and methods based on seed, seedling and plant characters, biochemical techniques namely electrophoresis of proteins and isoenzymes and DNA fingerprinting. International Seed Testing Association (ISTA), its role in development of seed testing procedures, rules and seed quality assurance for international seed trade.

Unit 5 : Seed Storage

Requirements and types of seed storage. Factors affecting seed storage and role of moisture, temperature, RH and moisture equilibrium. Viability nomographs. Seed deterioration causes and methods of control. Physiological, biochemical and molecular changes in seed ageing. Seed drying and Packaging needs. Storage structures. Methods of stacking and their impact. Short and medium term storage. Controlled storage. Germplasm storage. Cryopreservation. Design features of short, medium and long-term seed storage buildings. Operation and management of seed stores.

Unit 6 : Seed Health

Significance of seed health. Mode and mechanism of transmission of micro-organisms- fungi, bacteria and viruses. Procedures for seed health test and rules. Externally and internally seed - borne pathogens, mode of infection, development and spread, methods of detection of seed borne diseases. Important seed-borne diseases of cereals, oilseeds, pulses, fibre crops, vegetables and their control measures. Quarantine and International procedures of phyto-sanitary certificates. Important storage pests, their identification, monitoring and detection.ET value, nature and extent of damage, natural enemies and management. Use of pesticides, botanicals, mycotoxins for seed treatments. Carry over infestation, principles of fumigation and safe use of fumigants.

Unit 7: Seed Industry Development and Marketing

Trends in National and International seed industry development. International Seed Trade Federation (ISF) and Indian seed associations. Economics of seed production. Market survey, demand forecasting, pricing policies, marketing channels, planning and sales promotion. Buyer behavior and role of Government, semi Government, cooperative and private sectors in seed trade. Responsibilities of seed companies and dealers in Seed Act. Seed import and export.

Unit 8 : Protection of Plant Varieties

Plant Variety Protection (PVP) and its significance. Protection of Plant Varieties and Farmer's Right Act, 2001, its essential features. International Union for the Protection of New Varieties of Plants (UPOV) and its role in development of Plant breeders Rights and Seed Industry Development. Impact of PVP on seed supply system. DUS testing principles and application. Biodiversity Act. Criteria for protection of Essentially Derived Varieties (EDVs) and Genetically modified (GM) varieties.

ICAR-SRF Syllabus
Genetics

Section 1 : Basic Genetic Principles
Mendelian inheritance. Structure, organization and division of cells. Physical basis of heredity. Sex linkage and sex-determination. Linkage and crossing over, detection and estimation of linkage. Gene mapping in prokaryotes and eukaryotes. Gene concept, allelism and fine structure of the gene, Extra-chromosomal inheritance, male sterility and incompatibility.

Section 2 : Genetic Material
Structure, function, replication and repair. Primary and secondary structures of DNA and RNA, Genetic code. Gene enzyme relationships. Transcription and RNA processing. Translation. Gene regulation in prokaryotes and eukaryotes. Dynamic genome—insertion sequences, transposable elements, genome organization and repetitive DNA. DNA restriction and modification. Chloroplast, and mitochondrial genomes.

Section 3 : Chromosomes
Structure, function and replication. Chromosomal differentiation and banding techniques. Chromosome labeling and in situ hybridization. Karyotype. Artificial chromosomes. Special types of chromosomes.

Section 4 : Structural and Numerical Variations of Chromosomes
Interchanges, inversions, duplications and deletions. Polyploids, haploids, aneuploids and their utility. Chromosomal manipulation. Cytogenetic techniques for gene location and gene transfer. Wide hybridization.

Section 5 : Genome Analysis and Cytogenetics of Crop Plants
Wheat, maize, rice, Brassica, cotton, Vigna, potato and sugarcane.

Section 6 : Mutation
Induction, detection and utilization of mutation. Mutagens. Molecu-

lar mechanisms of mutation. Directed mutagenesis, Transposon-induced mutations.

Section 7 : Genetics of Development

Genome constancy in development. Nucleo-cytoplasmic interaction. Differential gene expression. Gene amplification. Cell determination, tissue differentiation and pattern formation. Homeotic genes. Double fertilization. Embryo and endosperm development. Seed storage protein synthesis. Human genetic disorders.

Section 8 : Genetic Engineering and Biotechnology

Artificial synthesis of gene. Gene cloning and vectors. Genomic and c-DNA library. Transposon tagging. DNA sequencing. Nucleic acid hybridization and immune-chemical detection. Chromosome walking

RFLP, RAPD and PCR approaches. Molecular markers for crop improvement. Recombinant DNA technology. Antisense RNA and ribozymes. Genetic transformation and transgenics.

Section 9 : Population Genetics

Gene and genotypic frequencies and forces changing gene frequency. Concept of natural, idealized and non-idealized populations. Hardy-Weinberg Law. Linkage disequilibrium. Small populations and genetic drift. Inbreeding and systems of inbreeding. Genetic load. Polymorphism.

Section 10 : Quantitative and Biometrical Genetics

Quantitative character. Multiple factor inheritance, polygenic variation. Concept of breeding value. Heritability. Selection response. Mating designs-diallel and line x tester. Combining ability, basic concepts and estimation of gene effects. Components of variance and their partitioning. Random and fixed effect models. Genotype x environmental interactions. Heterosis.

Section 11 : Statistics

Frequency distribution, Measures of central tendency. Probability theory and its implications in genetics, probability distributions and tests of significance. Correlation, linear, partial and multiple regression. Genetic divergence. D2 Design of experiments-basic principles, CRD, RBD and Split plot.

ICAR-SRF Syllabus
Plant Breeding

Section 1 : Role of Plant Breeding in Agriculture
Objectives of plant breeding. Development of plant breeding from ancient time. Early plant breeders and their accomplishments. Characteristics improved by plant breeding.

Section 2 : Reproductive Systems
Cell division, gametogenesis and fertilization. Sexual and asexual reproduction. Apomixis. Incompatibility and male sterility systems and their use in plant breeding.

Section 3 : Genetic Principles
Mendelian principles of inheritance. Linkage and crossing over. Mutation. Chromosomal aberrations and polyploidy. Extra chromosomal inheritance.

Section 4 : Quantitative Inheritance and Selection
Drift. Inbreeding, inbreeding depression and heterosis. Combining ability and heritability. Responses to selection, types of selection, and correlated response to selection. Mating systems. Estimation of variance components and covariance among relatives.

Section 5 : Breeding plans
Basic schemes: Pure lines and inbred lines. Breeding methods for improvement of self-pollinated plants, cross-pollinated plants and clones. Component and transgressive breeding. Back-crossing. Single-seed descent method. Multi-lines. Intra-and inter-population improvement methods. Development of synthetics and composites. Genetic basis of heterosis and development of hybrid cultivars. Concept of plant ideotype and its role in crop

improvement.

Section 6 : Genotype × Environment Interactions
Types of interactions. Assessment of genotype x environment interactions. Selection of testing locations. Allocations of resources. Stability of genotype performance and its estimation.

Section 7 : Breeding for Biotic and Abiotic Stresses
Kinds of resistance. Mechanisms for disease and insect resistance. Breeding for specific resistance and general resistance. Breeding for quality traits and for resistance to heat, frost, flood, drought and soil stresses.

Section 8: Special Techniques for Plant Improvement
Use of cytogenetical and biotechnological tools and techniques-haploids, aneuploids, wide-hybridization, embryo culture, meristem culture, cell and tissue culture, somaclonal variation, protoplast fusion, RFLP, RAPD, QTL and molecular marker assisted selection, gene tagging and antisense RNA technology.

Section 9 : Genetic Resources and Germplasm Conservation
Centers of origin, New domestication and new adaptation. Need for conservation strategy. Methods of maintenance.

Section 10 : Release, Seed Production and Distribution of Cultivars
Objectives and types of trials. Design of trials and trial systems. Breeder Seed multiplication. Hybrid seed production. production of parental lines. Identification and release of varieties. Seed certification and regulatory agencies. Plant Breeder's right and variety protection.

Section 11 : Statistical Methods and Field Plot Techniques
Sources of variation. Field plot techniques and methods for increasing accuracy of field experiments. Design of experiments and analysis of variance. Factorial schemes. Complete and incomplete block design, augmented design, grid and honeycomb designs. Hill plots, unreplicated evaluation. Data collection and interpretation. Correlation and regression. Tests of significance (t, F and chi-square).

ICAR-SRF Syllabus
Seed Science and Technology

Section 1 : Seed Biology

Floral biology, mode of reproduction, sporogenesis, pollination, fertilization, embryogenesis, fruit and seed development. Apomixis, parthenocarpy, polyembryony and somatic embryoids. Seed structure of monocot and dicot. Chemical composition of seeds. Seed dormancy—types, causes, mechanisms in induction and release, factors affecting, methods to overcome, bud dormancy, and significance in agriculture. Seed germination-requirements, imbibition pattern, physiological and biochemical changes, and role of hormones. Seed invigoration procedures.

Section 2 : Seed Production

Introduction to crop breeding methods. Variety testing, release and notification. Genetic purity, concept, and factors responsible for deterioration of varieties. Maintenance breeding. Generation system of seed multiplication. Seed production agencies. Identification of seed production areas and factors affecting. Compact area approach in seed production. Seed production planning, equipment, input and manpower requirement. Factors affecting pollination and seed set—temperature, humidity, wind velocity, insect pollinators, and supplementary pollination. Male sterility, self incompatibility and their role in hybrid seed production. Principles and methods of seed production of varieties and hybrids of cereals like wheat, paddy, sorghum, pearl millet and maize; pulses like chickpea, pigeon pea, green gram, black gram, soybean and cowpea; oilseeds like groundnut, brassica, sesame, sunflower and castor; fibre crops like cotton and Jute; vegetable crops like tomato, brinjal, okra, chilli, important cole and cucurbitaceous crops; important forages legumes and grasses, and true seed production in potato with reference to land requirements, isolation, roguing, seed crops management, time of harvesting, threshing/extraction methods. Seed production technology of plantation crops like coffee, tea, rubber, cocoa, cardamom and pepper. Disease free clonal propagation of crops like potato, sugarcane, sweet potato, tapioca, colocasia, betelvine; fruit crops like mango, citrus, banana, guava, sapota, pineapple, grape, apple, pear, plum, peach, apricot, tea rose; and seed production and clonal propagation of annual and perennial flowers like rose, gladiolus, chrysanthemum, marigold, dahlia, flex and petunia. Clonal standards and degeneration.

Section 3 : Seed Processing

Principles of seed processing. Seed drying—principles and methods Pre-cleaning, grading, treating, and packaging. Seed processing machines like cleaner-cum-grader, specific gravity separator, indented cylinder; seed treater, weighing and bagging machines, their operation and maintenance. Seed quality maintenance during processing.

Section 4 : Seed Quality Control

Seed legislation-Seed Act, seed rules, Seed Control Order and seed law enforcement. Seed certification history, concept, organization, phases and minimum certification standards. Field inspection principles and methods. Inspection at harvesting threshing and processing. Pre- and post- quality testing for genetic purity. Seed testing—concepts and objectives, role in seed quality control, seed sampling, seed moisture testing, purity analysis, germination testing, tolerance tests and equipments. Testing of treated and pelleted seeds. Quick viability tests. Seed vigour—concept, significance in productivity and storage, seed vigour tests. Testing for genuineness of varieties-principles and methods based on seed, seedling and plant characters, biochemical techniques namely electrophoresis of proteins and isoenzymes and DNA finger printing.

Section 5 : Seed Storage

Need for storage. Storage behavior of orthodox, recalcitrant and intermediary seeds. Factors affecting seed storage and role of moisture, temperature, RH, and moisture equilibrium. Seed deterioration causes, theories and methods of control. Packaging materials and hermetic packing. Storage structures. Methods of stacking and their impact. Short- and medium term storage. Controlled storage. Germplasm storage. Cryopreservation. Design features of short-medium-and long-term seed storage buildings. Operation and management of seed stores.

Section 6 : Seed Health

Significance of seed health. Mode and mechanism of transmission of micro-organisms-fungi, bacteria and viruses. Procedures for seed health test and rules. Externally and internally seed-borne pathogens, mode of infection, development and spread. Methods of detection of seed -borne diseases. Important seed-borne diseases of cereals, oilseeds, pulses, fibre crops, vegetables and their control measures. Use of fungicides,
botanicals, mycotoxins and their harmful effects. Quarantine and international procedures for phytosanitary certificate. Important storage pests, their identification, monitoring and detection, biology, ET value, nature and extent of damage, natural enemies and management. Important insecticides and their uses. Carry over infestation. Ecological factors. Principles of fumigation and safe use of fumigants. Botanicals for seed treatment.

Section 7 : Seed Marketing

Economics of seed production. Market survey, demand forecasting, pricing policies: marketing channels, planning and sales production. Buyer behavior and role of Government, semi-Government, co-operative and private sectors in seed trade. Responsibilities of seed companies and dealers in Seed Act. Seed import and export.

CONTENTS

1

Chronology

1. The chambers observed in cut section of cork were termed as cell by which one of the following?
(a) Leeuwenhoek (1674) (b) Robert Hooke (1665)
(c) Schwan (1839) (d) None of these
2. Chiasma was first discovered by which one of the following?
(a) Bateson and Punnett (1906) (b) Haldane (1942)
(c) Janssens (1909) (d) Morgan (1910)
3. Which one of the following scientists was first reported to the duplication?
(a) C. B. Bridges (1919) (b) Sturtevant (1926)
(c) T. H. Morgan (1910) (d) None of these
4. A chromosome is an organized structure of DNA, protein, and RNA found in cells. Which one of the following scientist coined the term chromosome?
(a) Balbiani (1881) (b) Darlington (1937)
(c) Strausberger (1875) (d) Waldeyer (1888)
5. Which one of the following scientists were first discovered to the polytene chromosomes in salivary glands of dipteran insects?
(a) Balbiani (1881) (b) Darlington (1937)
(c) Strausberger (1875) (d) Waldeyer (1888)
6. In which one of the following year, Mendel's work on peas was published?
(a) 1856 (b) 1866 (c) 1876 (d) None of these
7. In which one of the following year Mendel work was rediscovered?
(a) 1876 (b) 1884 (c) 1900 (d) 1905

8. The Hardy–Weinberg law was proposed in which one of the following year, states that allele and genotype frequencies in a population remain constant from generation to generation in the absence of other evolutionary influences.
(a) 1900 (b) 1906 (c) 1908 (d) 1912

9. The term genetics was first used and coined by William Bateson in which one of the following year?
(a) 1904 (b) 1905 (c) 1906 (d) None of these

10. The ABO blood group in man was first discovered by which one of the following?
(a) Benzer (1955) (b) Green (1961)
(c) Lendsteiner (1900) (d) Oliver (1941)

11. Which one of the following scientists was first described to the reproductive organ of a plant (maize)?
(a) Camerarious (1694) (b) Gaertner (1830)
(c) Koelreutor (1790) (d) Linnaeus (1760)

12. Which one of the following was first noted to the entry of pollen tube into ovary?
(a) Amici (1823) (b) Strausberger (1884)
(c) Thomas Fairchild (1717) (d) None of these

13. In which one of the following year Norman E. Borlaug was awarded Nobel Prize?
(a) 1970 (b) 1975 (c) 1980 (d) 1985

14. Properly working compound microscope was constructed by which one of the following?
(a) Dutrochet (1824) (b) Janssen and Janssen (1590)
(c) Leeuwenhoek (1665) (d) Schlieden and Schwon (1846)

15. Cell – lineage theory was proposed by which one of the following?
(a) Dutrochet (1824) (b) Robert Hook (1865)
(c) Virchow (1853) (d) None of these

16. The term mitosis was proposed by which one of the following?
(a) Altman (1889) (b) Brown (1838)
(c) Flemming (1821) (d) None of these

17. Which one of the following scientist was first used the term chromosome?
(a) Altman (1889) (b) Brown (1838)
(c) McClung (1902) (d) Waldeyer (1888)

18. The chromosome theory of inheritance was proposed in 1915 by which one of the following scientist?
(a) Thomas Hunt Morgan (b) Gregor Mendel
(c) Hugo de Vries (d) Jacob Bronowski

19. Jean-Baptiste Lamarck, in which one of the following year proposed to the theory of inheritance of acquired characteristics?
(a) 1806 (b) 1807 (c) 1808 (d) 1809

20. August Weismann, in which one of the following year putted forth to the germplasm theory?
(a) 1881 (b) 1882 (c) 1883 (d) 1884

21. In 1961, which one of the following scientists put forth to the operon hypothesis?
(a) Beadle and Tatum (b) Jacob and Monod
(c) Nirenberg and Khorana (d) Watson and Crick

22. The Journal of genetics was founded by which one of the following?
(a) Bateson and Punnett (1901) (b) Mendel (1866)
(c) Morgan (1910) (d) R. A. Fisher (1918)

23. In 1906, which one of the following person given to the coupling and repulsion phase of linkage?
(a) Bateson and Punnett (b) H. J. Muller
(c) Hutchinson (d) T. H. Morgan

24. Which one of the following scientist coined the term interference and coincidence?
(a) Haldane (1942) (b) Janssens (1909)
(c) Morgan (1933) (d) Mullar (1916)

25. Which one of the following scientists were first used to the term polygenes?
(a) D. S. Falconer (1960) (b) K. Mather (1941)
(c) R. A. Fisher (1918) (d) Sewall Wright (1935)

26. By which one of the following scientist, genetic variance was first divided into additive, dominance and epistatic components?
(a) Hayman (1954) (b) K. Mather (1949)
(c) R. A. Fisher (1918) (d) Sewall Wright (1921)

27. Partitioning of genetic variance into additive and non-additive components was suggested by which one of the following scientist?
(a) D. S. Falconer (1960) (b) Mather and Jinks (1982)
(c) R. A. Fisher (1918) (d) Sewall Wright (1935)

28. Partitioning of phenotypic variance into heritable fixable, heritable non-fixable and non-heritable non-fixable was suggested by which one of the following scientist?
(a) Falconer (1960) (b) K. Mather (1949)
(c) Kempthorne (1957) (d) R. A. Fisher (1918)

29. Sex chromosomes were first discovered by which one of the following?
(a) C. B. Bridges (1922) (b) McLung (1902)
(c) T. H. Morgan (1933) (d) Wilson and Stevens (1905)

30. The path from the initial discovery of sex chromosomes in 1891 to an understanding of their true function was paved by the diligent efforts of multiple scientists over the course of many years. Sex chromosomes were first discovered in which one of the following?
(a) Drosophila (b) Garden Pea
(c) Grasshoppers (d) Maize

31. Wilson and Stevens (1905) discovered sex chromosomes in which one of the following?
(a) Drosophila (b) Grasshopers
(c) Melandrium (d) None of these

32. Who one of the following scientists proposed to the genic balance theory of sex determination?
(a) Bateson and Punnett (1906) (b) Bridges (1922)
(c) McLung (1902) (d) Morgan (1910)

33. The first case of cytoplasmic inheritance was reported by which one of the following in 1909?
(a) Caspari (b) Correns
(c) Renner (d) Sturtevant

34. The double helical structure of DNA was proposed by which one of the following scientists?
(a) Avery, MacLeod and McCarty (1944)
(b) Beadle and Tatum (1941)
(c) Griffith (1928) (d) Watson and Crick (1953)

35. DNA is the genetic material was first discovered by which one of the following scientists?
(a) Avery, MacLeod and McCarty (1944)
(b) Benzer (1955)
(c) Griffith (1928)
(d) Hershey and Chase (1951)

36. Which one of the following scientists was first used to the term nucleic acid?
(a) Altmann (1889) (b) Mendel (1866)
(c) Miescher (1871) (d) Watson and Crick (1953)

37. Which one of the following scientists was proposed to the rolling circle model for replication of circular DNA?
(a) Cairns (1963) (b) Dressler (1968)
(c) Taylor (1969) (d) Watson and Crick (1953)

38. Which one of the following scientists was proposed to the 'one gene one enzyme' hypothesis?
(a) Beadle and Ephrussi (1937) (b) Beadle and Tatum (1941)
(c) Brenner (1955) (d) Crick (1966)

39. Reverse transcription generates in the cytoplasm as the process of synthesizing DNA using RNA as a template and reverse transcriptase as a catalyst. Which one of the following scientists was first report to the reverse transcription?
(a) Crick (1966) (b) Temin and Baltimore (1970)
(c) Watson and Crick (1953) (d) None of these

40. In 1941, Temin and Baltimore were awarded Nobel Prize for the discovery of which one of the following process?
(a) Reverse mutation (b) Reverse transcription
(c) Transcription (d) Translation

41. In 1961, which one of the following scientists was discovered the 'operon model of gene regulation in E. coli?
(a) Britten and Davidson (b) Jacob and Monod
(c) Temin and Baltimore (d) Watson and Crick

42. Nobel Prize for the discovery of operon model was awarded to Jacob and Monod in which one of the following years?
(a) 1960 (b) 1961 (c) 1961 (d) 1962

43. The term mutation was coined by which one of the following?
(a) Hugo de Vries (1900) (b) Morgan (1910)
(c) Muller (1927) (d) Stadler (1928)

44. For experimental induction of mutations in Drosophila by X-rays, the Nobel Prize was awarded to which one of the following?
(a) H. J. Muller (1927) (b) Hugo de Vries (1900)
(c) Stadler (1928) (d) T. H. Morgan (1910)

45. Which one of the following was first discovered mutation?
(a) Hugo de Viries (1900) (b) Muller (1927)
(c) Stadler (1928). (d) Wright (1971)

46. Pure-line theory provides the genetic basis for individual plant selection. In which one of the following years, pure-line theory was proposed by Johannsen?
(a) 1901 (b) 1902 (c) 1903 (d) 1904

47. By a selection procedure, single seed descent (SS(d) method is a derivation of plants in which F2 plants and their progeny are advanced by single seeds until genetic purity is achieved. Which one of the following was proposed to SSD method?
(a) Goulden (1939) (b) Grafius (1965)
(c) Marshall (1977) (d) Priestly (1970)

48. Which one of the following was scientist later modified to single seed descent (SS(d) method as a modification of the bulk breeding scheme?
(a) Goulden (1939) (b) Grafius (1965)
(c) Marshall (1977) (d) Priestly (1970)

49. For improvement in self-pollinated crops, a population approach method was suggested by which one of the following scientists?
(a) Goulden (1960) (b) Hull (1945)
(c) Janssen (1940) (d) Palmer (1953)

50. Who one of the following scientists reported to the male sterility in flowering plants?
(a) Fairchild (1717) (b) Gaertner (1730)
(c) Koelreuter (1763) (d) Linnaeus (1760)

51. Who one of the following scientist coined to the term selection?
(a) Darwin (1858) (b) Johannsen (1909)
(c) Mendel (1887) (d) None of these

52. Mass pedigree method is a modified pedigree method. Mass pedigree method of handling segregating generations was proposed by which one of the following scientists?
(a) Harrington (1937) (b) Jenkins (1940)
(c) Nilsson-Ehle (1945) (d) Sprague and Brimhall (1926)

53. Gamete selection was proposed by which one of the following scientists?
(a) Lonniquest (1952) (b) Richey (1927)
(c) Stadler (1944) (d) Vilmorin (1926)

54. The plants species bears sperm and eggs, both on the same gametophyte are known as Monoicous. This term was coined by which one of the following scientist?
(a) Strasburg's (1878) (b) Guignard (1899)
(c) Sewall wrignt (1921) (d) Darwin (1877)

55. In which one of the following year Germplasm theory was first proposed by August Weismann, states that inheritance is only affected by germ cells or gametes such as the egg and sperm cells?
(a) 1883 (b) 1883 (c) 1884 (d) None of these

56. Which one of the following people identified to the self and cross pollinated plants?
(a) Norman Borlaug (1965) (b) Sprengel (1800)
(c) Thomas Fairchild (1717) (d) None of these

57. Which one of the following scientists was developed to the semi-dwarf varieties of wheat?
(a) B. P. Paul (1960) (b) C. T. Patel (1970)
(c) Fraley (1983) (d) Norman Borlaug (1965)

58. Which one of the following scientists was originally developed to the concept of diallel selective mating system in 1970?
(a) Jenson (b) Johannsen
(c) Yule (d) None of these

59. Which one of the following people concerned with coining the term apomixes?
(a) Druery and Bower (1886) (b) Owen (1849)
(c) Winkler (1906) (d) None of these

60. Recurrent selection was divided by which one of the following, into two categories viz. intra-population improvement and inter-population improvement methods?
(a) Harvey (1960) (b) Moll and Sluber(1974)
(c) Richey (1926) (d) Stadler (1944)

61. In which one of the following year, Wilhelm Johannsen proposed to the genotype-phenotype distinction to make clear the difference between an organism's heredity and what that heredity produces?
(a) 1901 (b) 1910
(c) 1911 (d) None of these

62. Who one of the following scientists regarded as the 'Father of Hybrid Cotton' by developed the world's first ever commercial cotton hybrid, known as Hybrid-4 (Sankar-4) in 1970?
(a) Norman E. Borlang (b) M. S. Swaminathan
(c) Chandrakant T. Patel (d) T. S. Venkatraman

63. Which one of the following people proposed to the clean crop and dirty crop approach?
(a) Jenkins (1959) (b) Jensen (1952)
(c) Marshall (1977) (d) Richey (1956)

64. Wilhelm Roux expressed an idea in which one of the following years that chromosomes are the carriers of inheritance?
(a) 1881 (b) 1882 (c) 1883 (d) 1884

65. The term gene describes the fundamental physical and functional units of heredity. This term is coined by Danish botanist Wilhelm Johannsen in which one of the following year?
(a) 1906 (b) 1907 (c) 1908 (d) 1909

66. The parallel behavior of chromosomes and genes was established by which one of the following?
(a) Bateson and Punnett (1906) (b) Bridges (1922)
(c) Morgan (1910) (d) Sutton and Boveri (1903)

67. Which one of the following scientists was subdivided to the gene into recon, muton and cistron?
(a) Benzer (1955) (b) Green and Green (1949)
(c) Morgan (1933) (d) Watson and Crick (1953)

68. Which one of the following scientists was first discovered to the jumping gene?
(a) Barbara McClintock (1950) (b) Benzer (1955)
(c) Johannsen (1909) (d) Morgan (1933)

69. McClintock was awarded Nobel Prize in 1984 for the discovery of which one of the following?
(a) Jumping genes (b) Overlapping genes
(c) Pseudo genes (d) Split genes

70. Haemophilus influenzae was the first bacterial genome to be sequenced in which one of the following years?
(a) 1976 (b) 1995
(c) 2000 (a) None of these

71. In which one of the following years, Walter Fiers was established first sequenced RNA-genome Bacteriophage MS2 in virus?
(a) 1975 (b) 1976 (c) 1977 (d) 1978

72. Arabidopsis thalianawas was first plant genome sequenced in which one of the following years?
(a) 1999 (b) 2000
(c) 2001 (d) 2002

73. The botanist Hans Winkler was adapted term Genome in which one of the following year?
(a) 1919 (b) 1920
(c) 1921 (d) 1922

74. Which one of the following scientists were first discovered chromosomes?
(a) Benzer (1957) (b) Flemming (1821)
(c) Strausberger (1875) (d) Weismann (1882)

75. The term chromosome was coined by which one of the following?
(a) Benzer (1957) (b) Strausberger (1875)
(c) Waldeyer (1888) (d) Weismann (1882)

76. Which one of the following was coined to the term nucleotide?
(a) Benzer (1920) (b) Crick (1925)
(c) Levenc (1918) (d) Watson (1909)

77. The complete sequence of the circular genome in the human mitochondrion (16159 bp) was determined in which one of the following years?
(a) 1980 (b) 1981 (c) 1982 (d) 1983

78. Human genome project (HGP) is most recent achievement, in which one of the following years successful completions were of this project?
(a) 2001 (b) 2002 (c) 2003 (d) 2004

ANSWERSHEET

1. Chronology

1.	b	2.	c	3.	a	4.	d	5.	a	6.	b	7.	c
8.	c	9.	b	10.	c	11.	a	12.	a	13.	a	14.	b
15.	c	16.	c	17.	d	18.	a	19.	d	20.	c	21.	b
22.	a	23.	a	24.	d	25.	b	26.	c	27.	d	28.	b
29.	b	30.	c	31.	c	32.	b	33.	b	34.	d	35.	d
36.	a	37.	b	38.	b	39.	b	40.	b	41.	b	42.	b
43.	a	44.	a	45.	b	46.	c	47.	a	48.	b	49.	d
50.	a	51.	a	52.	a	53.	c	54.	c	55.	a	56.	b
57.	d	58.	a	59.	c	60.	b	61.	c	62.	c	63.	c
64.	c	65.	d	66.	d	67.	a	68.	a	69.	a	70.	b
71.	b	72.	b	73.	b	74.	c	75.	c	76.	c	77.	b
78.	b												

2

Cytogenetics

1. Which one of the following is a chromosome fragment lacking a centromere?
   (a) Acentric chromosome (b) Achrocentric chromosome
   (c) Telocentric chromosome (d) None of these
2. A chromosome or chromatid that has its centromere near the end is a modifying term for which one of the following?
   (a) Acentric (b) Acrocentric
   (c) Telocetric (d) All the above
3. A chromosome with several centromeres is called which one of the following?
   (a) Acentric (b) Dicentric
   (c) Monocentric (d) Polycentric
4. Which of the following statement is supporting to the process of meiosis, except?
   (a) Kinetochores are responsible for aligning chromatids during meiosis I.
   (b) Meiosis I separates chromosomes; meiosis II separates chromatids.
   (c) Segregation of unlinked alleles occurs during meiosis.
   (d) Synapsis and crossing-over occur during meiosis I.
5. In a chromosome, which one of the following one of a pair, or series, of alternative genes that occurs at a given locus?
   (a) A gene (b) Allele
   (c) Multiple allele (d) None of these
6. Which one of the following is such as different species is a polyploidy that has chromosome sets from different sources?
   (a) Allopolyploid (b) Aneuploid
   (c) Autopolyploid (d) None of these

7. In the F1 the hybrid species, which one of the following is termed for a species or type of plant derived from doubling of the chromosomes?
(a) Alloploids (b) Amphidiploid
(c) Autopolyploid (d) Both (a) and (b)

8. Toward the ends of the spindle, the daughter chromosomes pass from the equatorial plate to opposite poles of the cell. During which one of the following stages of mitosis or meiosis is
(a) Anaphase (b) Anaphase II
(c) Telophase (d) None of these

9. Monoploid which develop from a normal diploid species are referred to as which one of the following?
(a) Dihaploids (b) Disomic haploids
(c) Monohaploids (d) Polyhaploids

10. In the pairing of homologous chromosomes which one of the following is the failure or partial failure during the meiotic prophase is?
(a) Anticodon (b) Asynapsis
(c) tRNA (d) None of these

11. In the cell, which one of the following is an energy-rich compound that promotes their activities?
(a) ADP (b) ATP (c) DAP (d) All of the above

12. In Paramecium, which one of the following is a process of self-fertilization within one undivided cell, resulting in homozygosity occurs?
(a) Allogamy (b) Autogamy
(c) Syngamy (d) All of the above

13. Any chromosome but they are not considered as a sex chromosome. Which one of the following chromosomes is?
(a) Autosome (b) X chromosome
(c) Y chromosome (d) None of these

14. Which one of the following is a mutant bacterium (organism) requires the addition of some growth factor because they are not able to grow on a minimal medium?
(a) Auxotroph (b) Culture media
(c) Hetrotroph (d) All of the above

15. Which one of the following may result as part of a plant with a genetically different constitution as compared with other parts of the same plant, from different zygotes that grow together or artificial fusion (grafting)?
(a) Chimera (b) Clone (c) Legion (d) All of the above

16. In a plant cytoplasm, starch is synthesized in which one of the following green structures that contains chlorophyll?
(a) Chloroplastid (b) Chromoplast
(c) Leucoplast (d) Both (a) and (b)

17. A mode of cytoplasmic inheritance, independent of nuclear genes, has been associated with which one of the following cytoplasmic structures?
(a) Chloroplastid (b) Chromoplast
(c) Leucoplast (d) Both (a) and (b)

18. During mitosis or meiosis, which one of the following is one of the two identical strands resulting from self-duplication of a chromosome?
(a) Chromatid (b) Chromonema
(c) Chromosome (d) None of these

19. Which one of the following is/are a small body, identified by their characteristic viz. size and linear arrangement on the chromosome thread?
(a) Chromatin (b) Chromomeres
(c) Chromonema (d) Both (a) and (b)

20. Within each chromosome, which one of the following is an optically single thread forming an axial structure?
(a) Chromatid (b) Chromatin
(c) Chromonema (d) All of the above

21. The light and dark areas occur along the length of the chromosomes during staining. By which one of the following, chromosomes are identified?
(a) Chromosome banding (b) Heterochromatin
(c) Staining (d) None of these

22. During cell division, microscopically observable nucleoprotein bodies which are dark-staining with basic dyes in the cell. Which one of the following carries the genes that are arranged in linear order?
(a) Chromatids (b) Chromonema
(c) Chromosomes (d) All of the above

23. In the protein synthesis, a unit of function by which of the following that specifies one polypeptide chain in DNA is?
(a) Cistron (b) Muton
(c) Recon (d) None of these

24. Which one of the following is a termed for a set of nucleotide triplet that will code one amino acid is?
(a) Anticodon (b) Cistron (c) Codon (d) Muton

25. In messenger RNA, set of three adjacent nucleotides specifies the amino acid to be inserted in a specific position to form polypeptide during translation. Which one of the following is a correct termed?
(a) Anticodon (b) Codon
(c) Both (a) and (b) (d) None of these

26. Which one of the following is a coding triplet?
(a) Anticodon (b) Codon
(c) Both (a) and (b) (d) None of these

27. Which one of the following is a complementary resides in the structural gene specifying the mRNA question and also designations are listed in the amino acid and genetic code entries.
(a) Anticodon (b) Codon
(c) Both (a) and (b) (d) None of these

28. For the activity of an enzyme, which one of the following substances is very necessary?
(a) Coenzyme (b) Enzyme
(c) Inducer (d) Polypeptide

29. For transcription, less condensed form of eukaryotic chromatin that is a stain lighter in a karyotype is which one of the following?
(a) Euchromatin (b) Heterochromatin
(c) Both (a) and (b) (d) None of these

30. Transcriptionally inactive DNA mostly in repetitive DNA form of eukaryotic chromatin that appears darker in a karyotype?
(a) Euchromatin (b) Heterochromatin
(c) Both (a) and (b) (d) None of these

31. Which one of the following is an alkaloid is used to inhibit the formation of the spindle and delays the division of centromeres, derived from the Autumn crocus?
(a) Chenopodium (b) Colchicine
(c) Both (a) and (b) (d) None of these

32. Which one of the following an alkaloid that is used as an agent to stop mitosis at metaphase for preparation of Karyo types when chromosomes are maximally condensed?
(a) Asynapsis (b) Chenopodium
(c) Colchicine (d) All of the above

33. Which one of the following is used to produce polyploidy varieties of horticultural important species?

(a) Asynapsis (b) Chenopodium
(c) Colchicine (d) All of the above

34. The condition in linked inheritance in which an individual heterozygous for two pairs of genes received the two dominant members from one parent and the two recessives from the other parent is which one of the following?

(a) Coupling (b) Nondisjuction
(c) Repulsion (d) None of these

35. In two adjacent atoms, which one of the following bond having an electron pair is equally shared by protons?

(a) Covalent bond (b) Electron pair bond
(c) Both (a) and (b) (d) None of these

36. The exchange of genetic material between homologous chromosomes by breakage and reunion; it occurs during pairing of chromosomes at prophase-I of meiosis.

(a) Cross over suppressor (b) Crossing over
(c) Recon (d) Both (a) and (b)

37. A genetical process form new association of chromosomes which results in an exchange of genes and therefore produces new combinations different from its characteristics of the parents.

(a) Cross over suppressor (b) Crossing over
(c) Both (a) and (b) (d) None of these

38. A frequency of exchange of one percent of crossing-over value between two pairs of linked genes is equal to 1 unit on a linkage map is called as which one of the following?

(a) 2 centimorgon (b) Chiasmata
(c) Crossover unit (d) None of these

39. Which one of the following is area of biology concerned with the study of the organization of chromosomes in the nucleus with the requirements of gene expression in different developmental and environmental contexts?

(a) Cytogenetics (b) Cytology
(c) Molecular genetics (d) None of these

40. Which one of the following scientific discipline that deals with study of the behavior of chromosomes with transmission of phenotypes to progeny.

(a) Cytogenetics (b) Cytology
(c) Molecular genetics (d) None of these

41. Which one of the following are a scientific study of changes in chromosomal structure and number with speciation; and the evolution of the genome with the evolution of the species gene.
(a) Cytogenetics (b) Cytology
(c) Molecular genetics (d) None of these

42. Which one of the following in genetics, deals with study of cell chromosomes and their implications?
(a) Cytogenetics (b) Cytology
(c) Molecular genetics (d) Both (a) and (b)

43. Which one of the following is a part of mitosis or meiosis where cytoplasmic division and other changes exclusive of nuclear division?
(a) Cytokinesis (b) Karyokinesis
(c) Nuclear fission (d) None of these

44. Which one of the following is the study of the structure, function, and life history of the cell?
(a) Cytogenetics (b) Cytology
(c) Genetics (d) Molecular genetics

45. Which one of the following is the part of a cell, enclosed by the plasma membrane but excluding the nucleus?
(a) Cytoplasm (b) Endoplasmic reticulum
(c) Both (a) and (b) (d) None of these

46. The cell protoplasm outside the nucleus in which cell organelles are embedded is called as which one of the following?
(a) Cytoplasm (b) Endoplasmic reticulum
(c) Mitochondria (d) None of these

47. In the cytoplasm, which one of the following hereditary transmission depends on the cytoplasm or structures rather than the nuclear genes is?
(a) Cytoplasmic inheritance (b) Maternal inheritance
(c) Both (a) and (b) (d) None of these

48. The phenotypic differences found between individuals of identical genotype due to an effect of which one of the following inheritance?
(a) Biparental inheritance (b) Cytoplasmic inheritance
(c) Maternal inheritance (d) Uniparental inheritance

49. Reducing the number of loci by the absence of a segment or a band of a chromosome is termed as which one of the following?
(a) Deficiency or Deletion (b) Dublication
(c) Inversion (d) Translocation

50. The last stage in the prophase of meiosis-I, in which the bivalents are shortened and thickened, is known as which one of the following?
(a) Diakinesis (b) Leptotene
(c) Zygotene (d) None of these

51. Which one of the following is a chromosome with two centromeres?
(a) Dicentric chromosome (b) Polytene chromosome
(c) Both (a) and (b) (d) None of these

52. In which one of the following, the centres of the chromosome where sister chromatids are attached by a centromere?
(a) Metacentric (b) Submetacentric
(c) Telocentric (d) None of these

53. In which one of the following, just off-centre of the chromosome where sister chromatids are attached by a centromere?
(a) Metacentric (b) Submetacentric
(c) Telocentric (d) None of these

54. In which one of the following, at the end of the chromosome where sister chromatids are attached by a centromere?
(a) Metacentric (b) Submetacentric
(c) Telocentric (d) None of these

55. A diploid state is written as 2n to distinguish it from the haploid state (n) gene is known as which one of the following?
(a) Dimorphic (b) Diploid
(c) Disomic (d) None of these

56. Which one of the following is a short arm of the sister chromatids attached by a centromere?
(a) p arm (b) q arm
(c) Both (a) and (b) (d) None of these

57. Which one of the following is a long arm of the sister chromatids attached by a centromere?
(a) p arm (b) q arm
(c) Both (a) and (b) (d) None of these

58. In the prophase of first meiosis, the stage when the paired homologous chromosomes separate, except where they are held together by chiasmata?
(a) Diplotene (b) Pachytene
(c) Prophase (d) None of these

59. Which one of the following is a sub-stage of prophase of meiosis-I that preceding diakinesis but, follows the pachytene stage?
(a) Diplotene (b) Pachytene
(c) Prophase (d) None of these

60. Which one of the following is termed for movement of homologous chromosomes (of daughter chromosomes): toward the poles of the nuclear spindle during the separation in the anaphase of meiosis or mitosis division?
(a) Disjunction (b) Non-disjunction
(c) Terminalization (d) All the above

61. The separation of homologous chromosomes at the anaphase stage of meiosis or mitosis division is termed as which one of the following?
(a) Disjunction (b) Non-disjunction
(c) Terminalization (d) All the above

62. At the metaphase, failure separation of paired chromosomes which can occur both in meiosis and mitosis is?
(a) Disjunction (b) Non-disjunction
(c) Terminalization (d) All the above

63. A progressive shift of chiasmata from their sites of origin to more terminal positions is known as which one of the following?
(a) Disjunction (b) Non-disjunction
(c) Terminalization (d) All the above

64. In mitosis, which one of the following longest phase is?
(a) Anaphase (b) Metaphase
(c) Prophase (d) Telophase

65. More than one copy of a chromosomal segment within a chromosome set gene is termed as duplication.
(a) Deletion (b) Duplication
(c) Both (a) and (b) (d) None of these

66. The loss of a chromosomal segment from a chromosome set is which one of the following?
(a) Deletion (b) Duplication
(c) Both (a) and (b) (d) None of these

67. Inversion is a chromosomal aberration in which a block of genes is rotated by which one of the following?
(a) 120° (b) 160°
(c) 180° (d) 190°

68. On at least one locus gene, which one of the following is a diploid or polyploid individual that has different alleles?
(a) Embryo (b) Heterozygote
(c) Both (a) and (b) (d) None of these

69. Which one of the following is a female organism produce a germ cell?
(a) Embryo (b) Heterozygote
(c) Ovum (Egg) (d) All the above

70. Which one of the following is rudimentary plant that arises from the progressive differentiation in cell culture?
(a) Embryo (b) Heterozygote
(c) Both (a) and (b) (d) None of these

71. Which one of the following is a sister chromatids found near the end of the chromosome and attached by a centromere?
(a) Acrocentric (b) Cytogenetics
(c) Metacentric (d) Telocentric

72. Which one of the following is/are true statements about embryo?
(a) Germination in plants.
(b) Hatching in eutherian animals.
(c) Structure that develops from a zygote up to the time of birth.
(d) All of the above

73. Which one of the following is true statements regarding embryo sac?
(a) Female gametophyte of angiosperms, contain several haploid nuclei.
(b) The site of fertilization of the egg and development of the embryo.
(c) Formed by the division of the haploid megaspore nucleus.
(d) All of the above

74. A large thin-walled space within the ovule of the seed plant in which the egg and, after fertilization in higher plants, the embryo develops to the mature female gametophyte is termed as which one of the following?
(a) Embryo development (b) Embryo sac
(c) Ovum (d) All the above

75. Mitosis is the division of the nucleus of eukaryotic cells occurs in four stages designated as which one of the following order?
(a) Telophase. Prophase, Metaphase and Anaphase.
(b) Prophase, Metaphase, Anaphase and Telophase.
(c) Metaphase, Telophase. Anaphase and Prophase,
(d) Anaphase, Metaphase, Prophase and Telophase.

76. The centriole divides and the two daughter centrioles move apart during which one of the following?
(a) Anaphase (b) Metaphase
(c) Prophase (d) Telophase

77. The chromosomes move about within the spindle and eventually arrange themselves in the equatorial region of the spindle during which one of the following?
(a) Anaphase (b) Metaphase
(c) Prophase (d) Telophase

78. The block that prevented DNA replication in the centromere region is removed and the centromere becomes functionally double during which one of the following?
(a) Anaphase (b) Metaphase
(c) Prophase (d) Telophase

79. The spindle disappears and reconstruction of nuclear envelopes about the two groups of offspring chromosomes begins during which one of the following?
(a) Anaphase (b) Metaphase
(c) Prophase (d) Telophase

80. In which one of the following phase chromosomes becomes visible within the nucleus because they coil up to produce a series of compact gyres?.
(a) Anaphase (b) Metaphase
(c) Prophase (d) Telophase

81. Each chromosome is longitudinally double during which one of the following phases except in the region of the centromere?
(a) Anaphase (b) Metaphase
(c) Prophase (d) Telophase

82. Chromatid refers to each replicate strand of a chromosome during prophase. In which one of the following phase two chromatids are now ready to be separated and to move under the action of the traction fibres to the poles of the spindle?
(a) Metaphase (b) Prophase
(c) Both (a) and (b) (d) None of these

83. In which one of the following phases, the chromatids are converted to independent chromosomes that separate and move to opposite poles?
(a) Anaphase (b) Metaphase
(c) Telophase (d) None of these

84. As nuclear envelopes form around each group the chromosomes return of their extended state, and nucleoli reappear during which one of the following phases?
(a) Anaphase (b) Metaphase (c) Prophase (d) Telophase

85. Which one of the following, pairing of chromosomes or crossing-over between the homologous chromosomes does not involve?
(a) Anaphase of mitosis (b) Metaphase of mitosis
(c) Prophase of mitosis (d) Telophase of mitosis

86. A segment of the chromosome is A-B-C-D-E-F-G and the new chromosome is A-B-C-D-G refers to which one of the following?
(a) Chromosome rearrangement (b) Chromosome deletion
(c) Chromosome inversion (d) None of these

87. A segment of a chromosome is moved or copied somewhere else as 2n is termed as which one of the following is?
(a) Chromosome rearrangement (b) Paracentric inversion
(c) Both (a) and (b) (d) None of these

88. Which one of the following is a segment of the chromosome that 180° turned?
(a) Chromosome inversion (b) Chromosome rearrangement
(c) Chromosome translocation(d) None of these

89. Which one of the following is a segment of a chromosome moves to another place on the same chromosome or from one chromosome to a non-homologous chromosome?
(a) Chromosome inversion (b) Chromosome rearrangement
(c) Chromosome translocation (d) None of these

90. Which one of the following is true statements regarding mitosis?
(a) Chromatids of each chromosome separate and move to opposite poles at anaphase.
(b) Nuclear membrane develops around the chromosomes to define the cell nucleus.
(c) Two daughter nuclei are produced by nuclear division of mitosis.
(d) All of the above

91. Which one of the following process nucleus does not divide but chromosomes are replicate in a nucleus?
(a) Amitosis (b) Dublication
(c) Endomitosis (d) Mitosis

92. Which one of the following is correct for endomitosis?
(a) Endomitosis leads to endopolyploidy.
(b) Polytene chromosomes are formed in endomitosis
(c) Both (a) and (b) (d) None of these

93. Without division of the nucleus duplication of chromosomes, resulting in increased chromosomes within a cell where the chromosome strands separate but the cell does not divide is termed as which one of the following?
(a) Mitosis (b) Endomitosis
(c) Amitosis (d) Dublication

94. Which one of the following is a process of nuclear division other than mitosis?
(a) Endomitosis (b) Amitosis
(c) Dublication (d) None of these

95. Without participation of a mitotic apparatus, which one of the following is a term for the division of a nucleus into two parts by constriction?
(a) Amitosis (b) Endomitosis
(c) Both (a) and (b) (d) None of these

96. Which one of the following considered with amitosis is?
(a) A form of nuclear division in which a spindle is not formed.
(b) Nuclear membrane persists throughout the division, and the nucleus divides by constriction.
(c) Occurs in the macronuclei of ciliates and in certain fungi.
(d) All of the above

97. Which one of the following enzyme that breaks/cuts at specific positions to DNA strands which may involved in recombination of DNA is?
(a) Endonuclease (b) Restriction enzyme
(c) Both (a) and (b) (d) None of these

98. At internal positions of the strand rather than at the ends, which one of the following that catalyzes hydrolysis of a polynucleotide strand?
(a) Endonuclease (b) Hydrolysing agents
(c) Both (a) and (b) (d) None of these

99. **Ribosomes adhere by which one of the following network of membranes in the cytoplasm?**
(a) Endoplasmic reticulum (b) Mitochondria
(c) Plastids (d) Ribosome

100. **In diploid organisms, occurrence of cells is containing 4n, 8n, etc. like multiples 2n genomes. Which one of the following is?**
(a) Euploidy (b) Endopolyploidy
(c) Octoploid (d) Tetraploid

101. **Which one of the following considered with euploid is?**
(a) A cell and/or plant having any number of complete chromosome sets.
(b) Having eight chromosome sets of identical or different complements.
(c) Having four sets of chromosomes in the nucleus gene.
(d) None of these

102. **Which one of the following considered with octoploid is?**
(a) A cell and/or plant having any number of complete chromosome sets.
(b) Having eight chromosome sets of identical or different complements.
(c) Having four sets of chromosomes in the nucleus gene.
(d) None of these

103. **Which one of the following considered with tetraploid is?**
(a) A cell and/or plant having any number of complete chromosome sets.
(b) Having eight chromosome sets of identical or different complements.
(c) Having four sets of chromosomes in the nucleus gene.
(d) None of these

104. **Mitotic-type division (i.e. usually the non-reductional division) of meiosis II, the division in the meiotic sequence is consists of which one of the following?**
(a) Amitosis (b) Equatorial plate
(c) Polar ends (d) All of the above

105. **Which one of the following is estrus-producing compound and also known as female harmone?**
(a) Estrogen (b) Lectogen
(c) Prolactin (d) All of the above

106. Which one of the following genetic materials normally transcribed into mRNA but not stained by common dyes during interphase?
(a) Euchromatin (b) Heterochromatin
(c) tRNA (d) None of these

107. Euchromatin is chromatin or chromosome regions, which one of the following is considered with euchromatin?
(a) During interphase, relatively uncoiled and stains weakly.
(b) During metaphase, condensed and stains strongly.
(c) Shows normal staining properties of the bulk of the chro mosomal complement.
(d) All of the above

108. In the higher organisms, members are composed of cells with well developed nuclei which are enclosed in nuclear membranes. These organisms undergo meiosis are which one of the following?
(a) Bacterium (b) Eukaryote
(c) Prokaryote (d) None of these

109. Which one of the following terms is used to identify different levels in the series are diploid, triploid, tetraploid, and so on in an organism or cell having chromosome number that are an exact multiple of the monoploid (n) or haploid number?
(a) Aneukploid (b) Euploid
(c) Heteroploid (d) None of these

110. In the beginning at the ends of strand, which one of the following an enzyme that digests DNA or RNA is?
(a) Endonuclease (b) Exonuclease
(c) Both (a) and (b) (d) None of these

111. In the cytoplasm DNA units found that controls to the cytoplasmic inheritance. Which one of the following structure is but not a part of the nucleus?
(a) Extra nuclear chromosomes (b) Plastid
(c) Both (a) and (b) (d) None of these

112. A bacterium functions as a male in the presence of F+. Which one of the following fertility factor in a bacterium is?
(a) F factor (b) S factor
(c) Both (a) and (b) (d) None of these

113. Which one of the following process is used to form a zygote by fusion of a male gamete (sperm) with a female gamete (egg)?
(a) Autogamy (b) Fertilization
(c) Gametogenesis (d) All of the above

114. Which one of the following is considered with gametogenesis?
(a) Fertilizing of the nuclei of endosperm and embryo-forming cells by genetically different gametes.
(b) Obligatory self-fertilization e.g. pea, peanut, flax, barley, oats, lupine, rice, soybean, or wheat.
(c) The formation of gametes from gametocytes.
(d) All of the above

115. Which one of the following is considered with autogamy?
(a) Fertilizing of the nuclei of endosperm and embryo-forming cells by genetically different gametes.
(b) Obligatory self-fertilization e.g. pea, peanut, flax, barley, oats, lupine, rice, soybean, or wheat.
(c) The formation of gametes from gametocytes.
(d) All of the above

116. Which one of the following is considered with feterofertilization?
(a) Fertilizing of the nuclei of endosperm and embryo-forming cells by genetically different gametes.
(b) Obligatory self-fertilization e.g. pea, peanut, flax, barley, oats, lupine, rice, soybean, or wheat.
(c) The formation of gametes from gametocytes.
(d) All of the above

117. The cell is in resting or is not dividing in which one of the following stages of the cell cycle?
(a) Cytokynasis (b) Interphase
(c) Late telophase (d) Both (a) and (c)

118. Which one of the following is a unit of inheritance (DN(a) located at a fixed place on the chromosome and also specifying a biological function as hereditary determiner?
(a) Gene (b) Loci
(c) Locus (d) None of these

119. Which one of the following considered with the gene is?
(a) Hereditary unit, on the chromosome occupies a fixed position.
(b) May mutate to various allelic forms.
(c) Upon phenotype gene has a specific effect through transcription.
(d) All of the above

120. Which one of the following is a cell or an organism having only one genome or one complete set (n) of chromosomes?
(a) Diploid (b) Gamete
(c) Haploid (d) Both (a) and (b)

121. During Interphase, which one of the following chromatin staining darkly is?

(a) Chromatic (b) Euchromatin
(c) Heterochromatin (d) None of these

122. Producing unlike gametes with regard to the sex chromosomes, XY is male, the XX female

(a) Heterogametic sex (b) Homogametic sex
(c) Both (a) and (b) (d) None of these

123. A cell containing two or more different nuclei is termed as which one of the following?

(a) Chromatin (b) Heterokaryon
(c) Karyotype (d) None of these

124. Which one of the following is an organism containing a diploi3d number and characterized by an addition or a loss of the chromosome?

(a) Deletion (b) Dublication
(c) Euploid (d) Heteroploid or Aneuploid

125. Chromosomes occur in pairs and generally similar in size and shape but one comes to the male and other to the female parent. They contain the same gene loci and form bivalents at meiosis. Which one of the following is?

(a) Androsome (b) Homologous chromosomes
(c) Iso chromosomes (d) None of these

126. In a chromosome, which one of the following is termed for rearrangement of linear array of genes in such way that their order in the chromosome is reversed?

(a) Deletion (b) Duplication
(c) Inversion (d) Both (a) and (b)

127. A chromosome with two identical genes and identical arms. The arms are mirror images of each another known as which one of the following?

(a) Bivalent
(b) Iso chromosome
(c) Lampbrush chromosomes
(d) Polytene Chromosome

128. Strains organisms which are completely homozygous and genetically uniforms known as which one of the following?

(a) Cybrid (b) Hybrid
(c) Isogenic stocks (d) Both (a) and (b)

129. **Which one of the following considered with karyotype is?**
(a) Chromosome constitution of a cell or an individual.
(b) Chromosomes arranged in order of length and according to the position of centromere.
(c) Both (a) and (b) (d) None of these

130. **During mitotic metaphase which may be observed that entire chromosomal complement of an individual cell or individual is termed as which one of the following?**
(a) Genotype (b) Karyotype
(c) Phenotype (d) None of these

131. **The genetic constitution of an organism, which one of the following as opposed to its physical appearance (phenotype).**
(a) Genotype (b) Karyotype
(c) Both (a) and (b) (d) None of these

132. **The observable properties manifestated on an organism of a specific genotype produced in conjunction with the environment is/are which one of the following?**
(a) Karyotype (b) Phenotype
(c) Both (a) and (b) (d) None of these

133. **The chromosomes appear as single fine thread like structures but they are really double in meiosis Stage immediately preceding synapsis because DNA replication has taken place is which one of the following?**
(a) Leptotene (b) Prophase
(c) Zygotene (d) None of these

134. **Which one of the following is an allele for expression renders inviable cell or organism possessing it in proper arrangement?**
(a) Kappa particle (b) Lethal gene
(c) Sub lethal gene (d) All of the above

135. **Which one of the following is a given gene or one of its alleles occupies a fixed position on a chromosome?**
(a) Loci (b) Locus
(c) Both (a) and (b) (d) None of these

136. **By the destruction of the cell members Bursting of a bacterial cell following infection by bacteriophage is which one of the following?**
(a) Lethal gene (b) Lethality
(c) Lysis (d) All of the above

137. Which one of the following is/are.a large molecule; that has a high molecular weight as term used to identity molecules of proteins and nucleic acids?
(a) Macromolecule (b) Micromolecule
(c) Microtubles (d) None of these

138. Which one of the following term is used for a gene of the mother, controlled to the trait but expressed in the progeny?
(a) Eugenetics (b) Heterosis
(c) Maternal effect (d) Paternal effect

138. Which one of the following effect influence of the genotype or phenotype on the immediate offspring is?
(a) Maternal effect (b) Paternal effect
(c) Both (a) and (b) (d) None of these

140. Which one of the following considered with heterosis is/are?
(a) Always associated with increased heterozygosity;
(b) Increased vigor of growth, survival, and fertility of hybrids.
(c) Usually results from crosses between two genetically different, highly inbred lines.
(d) All of the above

141. For the genetic basis of heterosis, there are hypotheses have received the most attention is/are which one of the following?
(a) Dominance hypothesis (b) Overdominance hypothesis
(c) Both (a) and (b) (d) None of these

142. The chromosome number of a reproductive cell becomes reduced to half of somatic number or the diploid (2n) by which one of the following results in the formation of spores in plants is an important source of variability through recombination?
(a) Amitosis (b) Equatorial division
(c) Meiosis (d) All of the above

143. The chromosomes are most discrete and arranged on an equatorial plate in one of the among stage of the cell division which following prophase and preceding anaphase?
(a) Metaphors (b) Telophase
(c) Both (a) and (b) (d) None of these

144. Which one of the following component of the mitotic spindle is hollow filaments in the cytoplasm making up a part of the locomotor apparatuses of any motile cells?
(a) Endoplasmic reticulum (b) Microtubules
(c) Mitochondria (d) Plastids

145. From the DNA to the ribosomes for protein synthesis, which one of the following RNA that carries necessary information is/are?
(a) mRNA (b) rRNA
(c) tRNA (d) Both (a) and (c)

146. To the ends of the spindle and division of cytoplasm, which one of the following chromosome duplication followed by migration of chromosomes?
(a) Meiosis (b) Mitosis
(c) Both (a) and (b) (d) None of these

147. Which one of the following an organism or cell having one genome (n) or a single set of chromosomes is/are?
(a) Diploid (b) Haploid
(c) Momoploid (d) Both (b) and (c)

148. Lacking of one chromosome of proper complement (2n – 1) in a diploid organism is known as which one of the following?
(a) Monosomic (b) Nullisomic
(c) Tetrasomic (d) Trisomic

149. Which one of the following is/are with an unusually high mutation rate?
(a) Chromosome (b) Mutable genes
(c) Mutation (d) None of these

150. Either physical or chemical which one of the following an environmental agent capable for inducing mutations is?
(a) Mutant (b) Mutagen
(c) Mutation (d) All of the above

151. Mutation change brought about shows by an individual organism or a cell; a changed gene is which one of the following?
(a) Mutant (b) Mutagen
(c) Both (a) and (b) (d) None of these

152. In an organism, changes in the DNA at a particular locus are termed as which one of the following?
(a) Mutant (b) Mutagen
(c) Mutation (d) None of these

153. A constant mutation rate that adds mutant genes in a population which one of the following terms is/are used for repeated occurrences of mutations in a population
(a) Mutation pressure (b) Mutation rate
(c) Population pressure (d) All of the above

154. In mitosis or meiosis, too many chromosomes in some daughter cells and too few in others are the result of failure of separation or disjunction of homologous chromosomes known as which one of the following?
(a) Disjunction (b) Nondisjunction
(c) Both (a) and (b) (d) None of these

155. A chromosomal segment with genes that located at the secondary constriction of some metaphase chromosomes and control to the synthesis of ribosomal RNA is/are which one of the following?
(a) Nucleolar organizer (b) Primary constriction
(c) Ribosome (d) Both (a) and (c)

156. In the nucleus of metabolic cells, which one of the following is ribosome precursors, RNA-rich, spherical sack and associated with the nucleolar organizer?
(a) Mitochondria (b) Nucleolus
(c) Nucleus (d) Ribosome

157. Nucleic acid and protein are main component of conjugated protein. The chromosomes are made of which one of the following material?
(a) Lipoprotein (b) Nucleoprotein
(c) Ribose sugar (d) Both (b) and (c)

158. In eukaryotes, which one of the following spherical subunits, or "beads-on-a-string" of chromatin, that composed of a core particle and is measure approximately 100?
(a) Nucleosome (b) Ribosome
(c) Both (a) and (b) (d) None of these

159. Which one of the following is/are central part of a cell, surrounded by cytoplasm and containing genetic material?
(a) Mitochondria (b) Nucleolus
(c) Nucleus (d) Ribosome

160. Which one of the following is an organism or a diploid cell lacking both members of one specific pair represented as: 2n – 2 of chromosomes?
(a) Double nullisomics (b) Euploids
(c) Nullisomic (d) None of these

161. Which one of the following organism or cell with sets of chromosomes (8n) or eight genomes is?
(a) Aneuploid (b) Octoploid
(c) Pentaploid (d) Tetraploid

162. In meiosis, mid prophase stage immediately following zygotene and preceding diplotene. Which one of the following terms is used for visible as long, paired threads like chromosomes?
(a) Leptotene (b) Pachytene
(c) Zygotene (d) None of these

163. During the first meiotic division, which one of the following terms is used for first stage where chromosomes are appear as long, widely uncoiled, and single strands?
(a) Leptotene (b) Pachytene
(c) Zygotene (d) None of these

164. The DNA of each of the chromosomes has replicated during the first meiotic division where each chromosome consists of two identical members (chromatids) is which one of the following?
(a) Leptotene (b) Pachytene
(c) Zygotene (d) None of these

165. Which one of the following is/are considered with pachytene in the prophase state of the first meiosis?
(a) Crossing-over occurs between the non-sister chromatids of homologous chromosomes.
(b) Homologous chromosomes are completely paired.
(c) In heterozygous individual, recombination occurs during crossing-over.
(d) All of the above

166. Heterozygous individual is responsible for part of the segregation observed in the progeny considered with which one of the following?
(a) Leptotene (b) Pachytene
(c) Zygotene (d) None of these

167. The homologous chromosomes are associating side by side in the stage of first meiosis is known as synapse considered with which one of the following?
(a) Leptotene (b) Pachytene
(c) Zygotene (d) None of these

168. 2n = 20 chromosomes represents to a diploid crop species that form ten pairs of homologous chromosomes is/are which one of the following?
(a) Leptotene (b) Pachytene
(c) Zygotene (d) All of the above

169. In interphase, the giant chromosome cable-like formation consisting of many identical chromatids and produced by replication without division is known as which one of the following?
(a) Compound chromosome (b) Harlequin chromosomes
(c) Homologous chromosomes (d) Polytene chromosomes

170. In addition to the characteristic invariable complement of chromosomes. Which one of the following chromosomes presents often in varying numbers?
(a) Supernumerary chromosomes (b) Compound chromosome
(c) Homologous chromosomes (d) Accessory chromosome

171. Which one of the following is/are an intra-chromosomal inversion, does not include the centromere?
(a) Paracentric inversion (b) Pericentric inversion
(c) Both (a) and (b) (d) None of these

172. An intra-chromosomal inversion, which one of the following includes the centromere is/are?
(a) Paracentric inversion (b) Pericentric inversion
(c) Both (a) and (b) (d) None of these

173. Chromosomes in which sister chromatids stain that distinguish result one appears dark usually red-violet is?
(a) Alien chromosome (b) Compound chromosome
(c) Harlequin chromosomes (d) Supernumerary chromosomes

174. Which one of the following considered with alien chromosome is?
(a) A chromosome present in addition to the normal chromosome complement.
(b) A chromosome transferred from a more or less related species to a crop plant.
(c) Two separate chromosomes formed by the union.
(d) None of the above

175. Which one of the following considered with compound chromosome is?
(a) A chromosome present in addition to the normal chromosome complement.
(b) A chromosome transferred from a more or less related species to a crop plant.
(c) Two separate chromosomes formed by the union.
(d) None of the above

176. Which one of the following considered with accessory chromosome is?
(a) A chromosome present in addition to the normal chromosome complement.
(b) A chromosome transferred from a more or less related species to a crop plant.
(c) Two separate chromosomes formed by the union.
(d) None of the above

177. Which one of the following considered with nucleolar chromosome is/are?
(a) Also called as satellite chromosome.
(b) Carries the nucleolus organizer (region).
(c) Both (a) and (b) (d) None of these

178. Puffs are found in which one of the following chromosomes?
(a) A-chromosomes (b) B-chromosomes
(c) Lampbrush chromosome (d) Polytene chromosomes

179. Inversion leads to alteration in which one of the following?
(a) Composition of genes (b) Number of genes
(c) Sequence of genes (d) All of the above

180. Puffs are the sites of which one of the following synthesis?
(a) DNA synthesis (b) RNA synthesis
(c) Protein synthesis (d) None of these

181. Which one of the following replicated chromosomes consists after mitotic metaphase or anaphase II of meiosis is?
(a) Accessory chromosomes (b) Daughter chromosomes
(c) Homologous chromosomes (d) Supernumerary chromosomes

182. Which one of the following chromosomes has been observed in ovary, anthers, or immature seed tissues?
(a) Accessory chromosome (b) Compound chromosome
(c) Harlequin chromosomes (d) Polytene chromosomes

183. B chromosomes are also called as which one of the following?
(a) Accessory chromosomes (b) Extra chromosomes
(c) Supernumerary chromosomes (d) All of the above

184. In addition of morphology, B-chromosomes differ from A-chromosome by which one of the following is/are?
(a) Meiotic pairing (b) Mitotic behavior
(c) Numerical variation (d) All of the above

185. Which one of the following heterogeneous groups of chromosomes present in several plant species derives by aberrant division processes and subsequent modifications?
(a) A chromosome (b) B chromosome
(c) Both (a) and (b) (d) None of these

186. Which one of the following are the standard chromosomes of a given genome?
(a) A chromosome (b) B chromosome
(c) Both (a) and (b) (d) None of these

187. Between the interphase and metaphase stage of mitosis, the chromosome is longitudinally double except in the region of the centromere and each sister DNA strand from interphase becomes coiled. Which one of the following stage is/are?
(a) Metaphase (b) Prophase
(c) Telophase (d) None of these

188. The condition in linked inheritance in which an individual heterozygous for two pairs of linked genes receives the dominant member of one pair and the recessive member of the other pair from one parent and the reverse arrangement from the other parent (e.g., AAbb X aa B(b).
(a) Cis arrangement (b) Coupling
(c) Repulsion or Trans-arrangement (d) None of these

189. During meiosis, separation of the paternal from maternal chromosomes and consequent allele's separation and their phenotypic differences as observed in the offspring is termed as which one of the following?
(a) Independent assortment (b) Segregation
(c) Both (a) and (b) (d) None of these

190. Which one of the following is/are concerned with somatic cell?
(a) Referring to body cells/tissues, having two sets of chromosomes
(b) One set normally coming from the female and the other from the male parent are
(c) Both (a) and (b) (d) None of these

191. The occurrence of two loci on the same chromosome, which one of the following independent of distance that separates them is/are?
(a) Synapses (b) Synergamy
(c) Synteny (d) None of these

192. In interbreeding, natural populations that are reproductively isolated from which one of the following other such groups are?
(a) Class (b) Plants
(c) Species (d) All the above

193. Which one of the following cell undergoes a meiotic cycle (super mutagenesis) to form four haploid spermatids is?
(a) Sperm (b) Sperm mother cell
(c) Spexrmatids (d) All of the above

194. Maturation of the gametes (sperm) of the male takes place by which one of the following process?
(a) Oogenesis (b) Spermatids
(c) Spermatogenesis (d) None of these

195. Formation of sperm through spermatids, which one of the following is a part of spermatogenesis that follows the meiotic divisions of spermatocytes?
(a) Spermatids (b) Spermatogonia
(c) Spermiogenesis (d) All of the above

196. In the meiotic prophase, pairing of homologous chromosomes is/are known as which one of the following?
(a) Asynapsis (b) Desynapsis
(c) Synapsis (d) None of these

197. At the end of the first meiotic prophase, which one of the following ribbon like structure formed between synapsed homologs is/are?
(a) Dyad (b) Synaptinemal complex
(c) Tetrad (d) None of these

198. In plants, four cells arising from the second meiotic division. Which one of the following terms is/are also used to identify as quadruple of chromatids that is formed by the split of homologous chromosomes during meiosis?
(a) Terminalization (b) Tetrad
(c) Trisomics (d) None of these

199. In cell division, which one of the following is/are last stage where the chromosomes are moving towards the poles?
(a) Diakinesis (b) Prophase
(c) Telophase (d) None of these

200. Four haploid cells resulting from a single diploid cell during gametogenesis is known as which one of the following?
(a) Tetrad (b) Trisomics
(c) Both (a) and (b) (d) None of these

201. In genomes, which one of the following an organism whose cells contain four haploid (4n) sets of chromosomes is?
(a) Diploid (b) Octaploid
(c) Pentaploid (d) Tetraploid

202. Pertaining to an organism or a nucleus with four members of one of its chromosomes whereas the remainder of its chromosome complement is diploid known as which one of the following?
(a) Disomics (b) Nullisomics
(c) Tetrasomic (d) None of these

203. Which one of the following an undifferentiated cell (or nucleus) such as a blastomere that when isolated or suitably transplanted can develop into a complete embryo is/are?
(a) Gamete (b) Haploid
(c) Totipotent cell (d) None of these

204. Enzyme RNA polymerase catalyzes to the formation of RNA from ribonucleotide triphosphates by process through which messenger RNA is formed along a DNA template is/are termed as which one of the following?
(a) Transcription (b) Transformation
(c) Translation (d) None of these

205. In bacteria genetic recombination is mediated by bacteriophage is known as which one of the following?
(a) Transduction (b) Transformation
(c) Translation (d) None of these

206. From homozygous parent, offspring differing in three pairs of genes are which one of the following?
(a) Cybrid (b) Dihybrid
(c) Trihybrid (d) None of these

207. At meiosis, which one of the following is/are an unpaired chromosome?
(a) Bivalent (b) Chromotid
(c) Univalent (d) Both (a) and (b)

208. In the seed in certain plants, which one of the following is hereditary influence from the pollen parent to the endosperm?
(a) Cytoplasmic inheritance (b) Meta xenia
(c) Xenia (d) None of these

209. In reproduction the cell, which one of the following is/are produced by the union of two gametes to develop individual?
(a) Cybrid (b) Hybrid (c) Zygote (d) None of these

210. The credit for 'Cell theory' goes to which one of the following?
(a) Louis Pasteur and Teldon (b) Schlieden and Schwan
(c) Thomas fairchild (d) None of these

211. Credit for developing cell theory is usually given to which one of the following person?
(a) Matthias Jakob Schleiden (b) Rudolf Virchow
(c) Theodor Schwann (d) All of the above

212. Mitosis is also known as which one of the following?
(a) Heterotypic division (b) Homotypic division
(c) Reduction division (d) None of these

213. Meiosis is also referred to as which one of the following?
(a) Equational division (b) Homotypic division
(c) Reduction division (d) None of these

214. Meiosis gives rise to how many haploid/diploid cells?
(a) Two diploid cells (b) Two haploid cells
(c) Four haploid cells (d) None of these

215. In mitosis, daughter cells are identical to mother cell in which one of the following?
(a) Chromosome compliment (b) Shape
(c) Size (d) All of the above

216. The daughter cells produced by meiosis are different from mother cell in which one of the following?
(a) Shape and Size
(b) Chromosomes number and Chromosomes composition
(c) Both (a) and (c) (d) None of these

217. Synaptonemal complex develops during which one of the following in meiosis?
(a) Diplotene (b) Leptotene
(c) Pachytene (d) Zygotene

218. Chromosomes are arranged at equatorial plate during which one of the following in mitosis?
(a) Early metaphase (b) Early prophase
(c) Late metaphase (d) Late prophase

219. Duplication of chromosomes occurs in mitosis during which one of the following phase?
(a) Interphase (b) Metaphase
(c) Prophase (d) Telophase

220. B-chromosomes are found in which one of the following crops?
(a) Barley (b) Maize (c) Rice (d) Wheat

221. Sequence of sub-stages of prophase-1 is which one of the following in meiosis?
(a) Leptotene, Zygotene, Pachytene, Diplotene, Diakinesis
(b) Leptotene, Diplotene, Zygotene, Pachytene, Diakinesis
(c) Leptotene, Pachytene, Zygotene, Diakinesis, Diplotensis
(d) Zygotene, Leptotene, Pachytene, Diakinesis, Diplotene

222. A chromosome with diffused centromere is referred to as which one of the following?
(a) Acrocentric (b) Holokinetic
(c) Telocentric (d) None of these

223. Which one of the following is/are chromosome with centromere and two arms of about equal length near the middle?
(a) Metacentric chromosome (b) Telocentric chromosome
(c) Both (a) and (b) (d) None of these

224. A metacentric chromosome at anaphase will assume as which one of the following shape?
(a) V shape (b) J shape
(c) Rod shape (d) None of these

225. A chromosome with median position of centromere is termed as which one of the following?
(a) Acrocentric (b) Metacentric
(c) Sub metacentric (d) Telocentric

226. The movement of chromosomes at anaphase is associated with which one of the following?
(a) Centromere (b) Chromomere
(c) Telomere (d) None of these

227. Which one of the following represents to the monoploids?
(a) n (b) x (c) 2n (d) 2x

228. Match the following stages of mitosis occurs during cell division with their phenomenon.

1. Prophase	A. Chromatids separate at the centromere and migrate to opposite poles
2. Metaphase	B. The two new nuclei assume their nor mal structure, and cell division is com pleted, producing two new daughter cells.
3. Anaphase	C. Each chromosome consists of two chro matids joined at the centromere.
4. Telophase	D. Chromosomes align at the center of the cell.

(a) 1B 2C 3D 4A (c) 1C 2D 3A 4B
(b) 1C 2D 3B 4A (d) 1D 2C 3A 4B

ANSWERSHEET

2. Cytogenetics

1. a	2. b	3. d	4. a.	5. b	6. a	7. b
8. a	9. c	10. b	11. b	12. b	13. a	14. a
15. a	16. a	17. a	18. a	19. b	20. c	21. a
22. c	23. a	24. c	25. b	26. b	27. b	28. a
29. a	30. b	31. b	32. c	33. c	34. a	35. a
36. b	37. b	38. c	39. a	40. a	41. a	42. a
43. a	44. b	45. a	46. a	47. b	48. c	49. a
50. a	51. a	52. a	53. b	54. c	55. b	56. a
57. b	58. a	59. a	60. a	61. a	62. b	63. c
64. c	65. b	66. a	67. c	68. b	69. b	70. c
71. b	72. a	73. d	74. b	75. b	76. b	77. c
78. b	79. a	80. d	81. c	82. c	83. a	84. a
85. d	86. c	87. b	88. a	89. a	90. c	91. d
92. c	93. c	94. b	95. b	96. a	97. d	98. c
99. a	100. a	101. b	102. a	103. c	104. b	105. a
106. a	107. d	108. b	109. b	110. b	111. a	112. a
113. b	114. c	115. b	116. a	117. b	118. a	119. d
120. c	121. c	122. b	123. a	124. d	125. b	126. c
127. b	128. c	129. c	130. b	131. a	132. b	133. a
134. b	135. b	136. a	137. a	138. c	139. a	140. d
141. b	142. c	143. a	144. b	145. a	146. b	147. d
148. a	149. b	150. b	151. a	152. c	153. a	154. b
155. a	156. b	157. b	158. c	159. c	160. c	161 b
162. b	163. a	164. a	165. d	166. b	167. c	168. c

169. d	170. a	171. a	172. b	173. c	174. b	175. c
176. a	177. c	178. d	179. c	180. a	181. b	182. d
183. d	184. d	185. b	186. a	187. b	188. c	189. b
190. c	191. c	192. c	193. b	194. c	195. c	196. c
197. b	198. b	199. c	200. a	201. d	202. c	203. c
204. a	205. a	206. c	207. c	208. c	209. c	210. c
211. d	212. b	213. c	214. c	215. d	216. b	217. d
218. a	219. a	220. b	221. a	222. b	223. a	224. a
225. b	226. a	227. b	228. c			

3

Genetics

1. How many scientists were rediscovered independently to the Mendel work?
(a) One scientist (b) Two scientists
(c) Three scientists (d) Four scientists

2. Which one of the following group of scientists was discovered independently to the Mendel work?
(a) Correns, Naudin and Knight
(b) de Vries, Correns and Tschrmak
(c) Koelreutor, John Goss and Gaertner
(d) Naudin, Koelreutor and Gaertner

3. The Hardy–Weinberg law may be applied in which one of the following ways?
(a) A population is assumed to be in Hardy–Weinberg proportions, in which the genotype frequencies can be calculated.
(b) The genotype frequencies of all three genotypes are known, they can be tested for deviations that are statistically significant.
(c) Both (a) and (b) (d) None of these

4. Each type of chromosome is represented by which one of the following in a genome?
(a) Only once (b) Thrice
(c) Twice (d) Many times

5. Between which one of the following, Brassica juncea is an amphidiploid?
(a) B. Campestris X B. nigra (b) B. campestris X B. oleracea
(c) B. nigra X B. oleracea (d) None of these

6. Gametes are formed by differentiation of which of the following?
(a) Germplasm cells (b) Somatoplasm cells
(c) Both (a) and (b) (d) None of these

7. The book "Origin of species" was published in the year 1859, was proposed by which one of the following?
(a) Charles Darwin (b) Koelreutor
(c) Lamarck (d) Louis pasteur

8. The law of Homologues series of variation is one of the fundamental laws of genetics. This law was proposed in 1920 by which one of the following scientist?
(a) Dobzonsky (b) Kolreutor
(c) M. S. Swaminathan (d) N. I. Vavilov

9. ABO blood group is an example of which one of the following?
(a) Single alleles (b) Double alleles
(c) Multiple alleles (d) None of these

10. Existence of more than two alleles at a locus is referred to which one of the following terms?
(a) Isoalleles (b) Multiple alleles
(c) Pseudo alleles (d) None of these

11. Law of Homologous Series is relevant for the Future because which one of the following?
(a) Allows predictionsin systematics.
(b) Allows predictionsin actual plant breeding.
(c) Distinguishes between apparent and real plant.
(d) All of the above

12. Which one of the following person concerne with the evidences for spontaneous generation of micro organism?
(a) Anton van Leeuwenhoek (b) Lazzaro Spallanzani
(c) Louis pasteur (d) None of these

13. Theory of evolution was proposed by which one of the following persons?
(a) Charles Darwin (b) Koelreutor
(c) Lamarck (d) None of these

14. Which one of following is the basic fundamental law of population genetics?
(a) Hardy Weinburg law
(b) Law of independent assortment
(c) Law of multiple factor hypothesis
(d) Law of segregation proposed by Mendel

15. What would be the size of chromosome having duplication?
(a) Normal (b) Shorter than normal
(c) Longer than normal (d) Thicker than normal

16. Pseudo-dominance is associated with which one of the following?
(a) Deletions (b) Duplication
(c) Inversions (d) Translocations

17. The chromosome theory of inheritance was proposed by which one of the following?
(a) Sutton and Boveri (b) John Knight and John Goss
(c) Johnson and Johnson (d) Louis Pasteur

18. Dihybrid test cross ratio is which one of the following?
(a) 1:1:1:1 (b) 1:2:1:1
(c) 9:3:3:1 (d) None of these

19. Gregor Mendel conducted hybridisation experiments on which one of the following crop for seven years from 1856-1863.
(a) Garden peas (b) Green peas
(c) Both (a) and (b) (d) None of these

20. Which one of the following scientist proposed the laws of inheritance in living organisms?
(a) Gregor Mendel (b) Hugo de Vries
(c) Jacob Bronowski (d) William Bateson

21. Who is well known as the "father of modern genetics"?
(a) Gregor Mendel (b) Hugo de Vries
(c) Jacob Bronowski (d) William Bateson

22. Which of the following explain to the Mendel's Laws of Inheritance?
(a) Law of Segregation (b) Law of Independent Assortment
(c) Both (a) and (b) (d) None of these

23. During pachytene translocation heterozygotes lead to formation of which one of the following?
(a) Cross shaped configuration (b) Loop configuration
(c) Ring configuration (d) None of these

24. Deletion leads to alteration in which one of the following?
(a) Gene number (b) Sequence of gene
(c) Structure of gene (d) All of the above

25. Genotypes are made up of two of the same letter which may?
(a) Either two capital (b) Two lowercase
(c) One of each (d) All of the above

26. Which of the following is a possible abbreviation for a genotype?
(a) BC (b) fg (c) Pp (d) Ty
27. Which of the following pairs is not correct?
(a) Hybrid = heterozygous (b) Heterozygous = Hh
(c) Homozygous = RR (d) All of the above
28. The genes present in an organism represent the organism's ________.
(a) Genotype (b) Phenotype
(c) Physical traits (d) None of these
29. How many alleles for one trait are normally found in the genotype of an organism?
(a) 1 (b) 2 (c) 3 (d) 4
30. Which one of the following statement is not true?
(a) A phenotype is the physical appearance of a trait in an organism.
(b) Alleles are different forms of the same gene.
(c) Genotype determines phenotype.
(d) Phenotype determines genotype.
31. Mendel's Law of Segregation is best illustrated by which one of the following cross?
(a) Bb x Bb (b) Hh x hh
(c) rr x rr (d) TT x tt
32. In a certain plant, purple flowers are dominant to red flowers. If the cross of two purple-flowered plants produces some purple-flowered and some red-flowered plants. What is the genotype of the parent plants?
(a) PP x Pp (b) Pp x Pp
(c) pp x PP (d) pp x pp
33. A plant that has purple flowers is crossed with one that has white flowers. Which one of the following were the genotypes of the parent plants, if the offspring were half white and half purple?
(a) Pp x pp (b) Pp x Pp
(c) PP x pp (d) PP x WW
34. Crossing two dihybrid organism results in which one of the following phenotypic ratio?
(a) 1:1 (b) 1:2:1
(c) 3:1 (d) 9:3:3:1

35. A dihybrid cross (AaBb x AaB(b) will result in which one of the following offspring ratio?
(a) 1:3:1 (b) 3:1
(c) 9:3:3:1 (d) 4:4:2:2

36. In an organism, the outward appearance (gene expression) of a trait is referred as which one of the following?
(a) An allele (b) Genotype
(c) Independent assortment (d) Phenotype

37. Change in chromosome number which involves entire set of genome is known as which one of the following?
(a) Aneuploidy (b) Euploidy
(c) Trisomy (d) None of these

38. The extra chromosome is which one of the followingin primary trisomic?
(a) Isochromosome (b) Normal chromosome
(c) Traslocated chromosome (d)None of these

39. Mendel formulated his Law of Segregation after he had which one of the following?
(a) Produced hybrids (b) Produced mutations
(c) Studied F1 offspring (d) Studied F2 offspring

40. Law of Dominance is illustrated by which one of the following cross produce phenotypic ratios?
(a) TT x tt (b) TT x Tt
(c) Tt x Tt (d) tt x tt

41. Law of Segregation is illustrated by which one of the following cross produce phenotypic ratios?
(a) TT x tt (b) TT x Tt
(c) Tt x Tt (d) tt x tt

42. Which one of the following represents to the nullisomic individual?
(a) 2n-1 (b) 2n+1
(c) 2n-2 (d) 2n+2

43. The mating of two curly-haired brown guinea pigs results in some offspring with brown curly hair, some with brown straight hair, some with white curly hair, and even some with white straight hair. This mating illustrates which of Mendel's Laws?
(a) Dominance (b) Independent Assortment
(c) Segregation (d) Sex-Linkage

44. Which one of the following law used to explains the proportion of 3:1 obtained at the F2?
(a) Law of dominance (b) Law of independent assortment
(c) Law of segregation (d) None of these

45. Which one of the following statements is true regarding Mendel?
(a) He believed that genetic traits of parents will usually blend in their progeny.
(b) His discoveries concerning genetic inheritance were generally accepted by the scientific community when he published them during the mid-19th century.
(c) His ideas about genetics apply equally to plants and animals.
(d) All of the above

46. Mendel believed that the characteristics of pea plants are determined by which one of the following?
(a) Inheritance of units or factors from both parents
(b) Inheritance of units or factors from one parent
(c) Relative health of the parent plants at the time of pollination
(d) None of these

47. An allele is defined as which one of the following?
(a) A heterozygous genotype (b) A homozygous genotype
(c) Another word for a gene
(d) One of several possible forms of a gene

48. Phenotype refers to which one of the following of an individual?
(a) Actual physical appearance (b) Genetic makeup
(c) Recessive alleles (d) All of the above

49. When the genotype consists of a dominant and a recessive allele, the phenotype will be like which one of the following allele?
(a) The dominant (b) The recessive
(c) Both (a) and (b) (d) None of these

50. The idea that different pairs of alleles are passed to offspring independently is Mendel's principle of which one of the following?
(a) Independent assortment (b) Segregation
(c) Unit inheritance (d) None of these

51. The idea that for any particular trait, the pair of alleles of each parent separate and only one allele from each parent passes to an offspring is Mendel's principle of which one of the following?
(a) Hybridization (b) Independent assortment
(c) Segregation (d) None of these

52. Two true breeding parents are crossed similar to Mendel's P generation. Which one of the following is the expected outcome for the F1 generation, if a tall plant is crossed with a short plant?
(a) All medium height (b) All short
(c) All tall (d) Half tall, half short

53. Which one of the following is the expected outcome for the F1 generation, if the cross from #1 is continued?
(a) 3 tall, 1 short (b) All short
(c) All tall (d) Half tall, half short

54. Which one of the following is its genotype, if a plant that has round seeds (Assume that round is dominant) has a parent that has wrinkled seeds?
(a) RR (b) Rr
(c) rr (d) RrWw

55. During gamete formation, alleles are separate. This is stated by which one of the following principle?
(a) Principle of dominance and recessiveness
(b) Principle of independent assortment
(c) Principle of multiple alleles
(d) Principle of segregation

56. How many of the offspring will also be heterozygous, if crossed two heterozygous plants?
(a) 1/4 (b) 3/4
(c) All (d) Half

57. Which one of the following principles of genetics supports to the observation: 'in each case where Mendel crossed true breeding plants as parents, the offspring displayed only one of the two traits seen in the parents'?
(a) Law of dominance and recessiveness
(b) Law of independent assortment
(c) Law of segregation (d) Allele frequency

58. The cross tall, round (TTRR) x short, wrinkled (ttrr); which one of the following outcomes is expected?
(a) All offspring short and wrinkled
(b) All offspring tall and round
(c) Half tall, round; half short wrinkled (d) None of these

59. Which one of the following theories illustrates to the development involves differentiation of an initially undifferentiated entity.
(a) Preformation theory (b) The germplasm theory
(c) Theory of epigenesis (d) Theory of pangenesis

60. Theory of epigenesis was proposed by which one of the following scientists?
(a) August Weismann (b) Caspar Friedrich Wolff
(c) Hugo de Vries (d) None of these

61. Which one of the following is a theory of heredity in which gemmules containing hereditary information from every part of the body and are incorporated into the reproductive cells?
(a) Preformation theory (b) The germplasm theory
(c) Theory of epigenesis (d) Theory of pangenesis

62. Which one of the following scientists proposed to the theory of pangenesis?
(a) August Weismann (b) Charles Darwin
(c) Hugo de Vries (d) None of these

63. Which one of the following theories illustrate that physiological modifications acquired by an organism can be inherited by the offspring?
(a) Preformation theory (b) The germplasm theory
(c) Theory of acquired character (d) Theory of epigenesis

64. Which one of the following scientists put forth to the theory of inheritance of acquired characteristics?
(a) Charles Darwin (b) Gregor Mendel
(c) Hugo de Vries (d) Jean-Baptiste Lamarck

65. Auto polyploidy is referred to which one of the following?
(a) Hybrid polyploidy (b) Segmental polyploidy
(c) Simple polyploidy (d) None of these

66. Which one of the following is a theory of the physical basis of heredity illustrate that inheritance is only affected by "germ cells" or gametes such as the egg and sperm cells?
(a) Preformation theory (b) The germplasm theory
(c) Theory of acquired character (d) Theory of epigenesis

67. Which one of the following scientists advocated to the germplasm theory?
(a) August Weismann (b) Charles Darwin
(c) Hugo de Vries (d) Jean-Baptiste Lamarck

68. Translocation homozygotes lead to alteration in which one of the following?
(a) Gene number (b) Linkage groups
(c) Linkage map (d) Sequence of genes

69. Which one of the following theories illustrate that an individual develops by simple enlargement of a tiny fully formed organism that exists in the germ cell?
(a) The germplasm theory (b) Theory of acquired character
(c) Theory of epigenesist (d) Theory of preformation

70. Which one of the following scientists proposed to the preformation germplasm theory?
(a) Bateson and Punnett (b) Lamarck and Weismann
(c) Swammerdam and Bonnet (d) None of these

71. Manifold effects of a gene refer to which one of the following?
(a) Penetrance (b) Expressivity
(c) Pleiotropy (d) Epistasis

72. The situation in which one gene makes the expression of another means to say nonreciprocal interaction of non-allelic genes is termed as which one of the following?
(a) Penetrance (b) Expressivity
(c) Pleiotropy (d) Epistasis

73. Which one of the following term is used for the phenomenon of a single gene being responsible for a number of different phenotypic effects that are apparently unrelated?
(a) Penetrance (b) Expressivity
(c) Pleiotropy (d) Epistasis

74. Between which one of the following, translocation leads to exchange of segments?
(a) Homologous chromosomes
(b) Non-homologous chromosomes
(c) Sister chromatids (d) None of these

75. The proportion of individuals of a specified genotype who manifest that genotype in the phenotype under a defined set of environmental conditions is known as which one of the following?
(a) Penetrance (b) Expressivity (c) Pleiotropy (d) Epistasis

76. In the phenotype, which one of the following term is used for the degree to which a particular genotype is expressed?
(a) Penetrance (b) Expressivity
(c) Pleiotropy (d) Epistasis

77. The genes are the hereditary units found within the nucleus of every living cell of an organism in the form of chromosomes, composed of which one of the following?
(a) DNA (deoxyribonucleic acid)
(b) RNA (deoxyribonucleic acid)
(c) Protein molecules (d) Both (a) and (c)

78. In endosperm of a diploid plant (2n=20), which one of the following chromosome number is?
(a) 10 (b) 20 (c) 30 (d) 40

79. Which one of the following true regarding the eukaryote is an organism?
(a) Whose cells have double nuclei
(b) Whose cells have extra nuclei.
(c) Whose cells have not true nuclei.
(d) Whose cells have true nuclei.

80. The centromere is also known as which one of the following?
(a) Chromomere (b) Primary constriction
(c) Secondary constriction (d) None of these

81. The F2 phenotypic ratio of a co-dominant trait governed by a single dominant gene is termed as which one of the following?
(a) 2:1 (b) 3:1 (c) 1:2:1 (d) None of these

82. Co-dominance is the phenomenon when________.
(a) One allele makes the effects of another allele.
(b) Both the allele of a gene expresses them in the heterozygote.
(c) Two dominant genes present together are capable of producing a character.
(d) All of the above

83. Intercalary deletion leads to loss of segment in a chromosome from the which one of the following?
(a) Intermediate position (b) Terminal portion
(c) Both (a) and (b) (d) None of these

84. With completed dominance and equal survival of all genotypes, the genes in F2 progeny will have phenotypic into which one of the following ratios?
(a) 1:2 ratio (b) 3:1 ratio
(c) 1:2:1 ratio (d) None of these

85. With complete dominance and equal survival of all genotypes, the genes in F2 in a monohybrid cross segregate into which one of the following ratios?
(a) 1:2 ratio (b) 3:1 ratio
(c) 1:2:1 ratio (d) None of these

86. Substitution haploids are represented as which one of the following?
(a) n – 1 (b) n +1
(c) n · 1 + 1 (d) None of these

87. In pea the gray seed coat color (G) is dominant over white (g). 82 gray and 78 white progeny are produced in a cross between gray and white, what is the probable genotype of the parents?
(a) Gg x gg (b) GG x gg
(c) Gg x Gg (d) None of these

88. In trihybrid cross, how many number of genotypes obtained?
(a) 3 (b) 6
(c) 9 (d) 64

89. 9:7 ratios are obtained in gene interaction of which one of the following?
(a) Complementary (b) Duplicate
(c) Inhibitory (d) Supplementary

90. In F2 the phenotypic ratio of 9:3:3:1 is modified to which one of the following in recessive epistasis?
(a) 9:7 (b) 15:1 (c) 9:3:4 (d) 12:3:1

91. Dominant epistasis is also known as which one of the following?
(a) Complementary epistasis (b) Duplicate epistasis.
(c) Simple epistasis (d) Supplementary epistasis

92. The gene which has masking effect In gene interaction is termed as which one of the following?
(a) Codominant gene (b) Dominant gene
(c) Epistatic gene (d) Recessive gene

93. Which one of the following term is used for the gene whose effect is masked in inter locus interaction?
(a) Dominant gene (b) Epistatic gene
(c) Hypostatic gene (d) Recessive gene

94. Tendency of one crossover to enhance the chance of another crossover in its adjacent region is referred to as which one of the following?
(a) Coincidence (b) Coupling
(c) Negative interference (d) Positive interference

95. Crossing over takes place during which one of the following?
(a) Diplotene (b) Leptotene
(c) Pachytene (d) Zygotene

96. Structural changes in chromosome cause alteration in which one of the following?
(a) Karyotype (b) Phenotype
(c) Viability (d) All of the above

97. Additive genes exhibit by which one of the following?
(a) Complete dominance (b) Incomplete dominance
(c) Lack of dominance (d) Over dominance

98. Additive genetic variance is which one of the following?
(a) Heritable fixable (b) Heritable non-fixable
(c) Non-heritable non fixable (d) None of these

99. Additive genetic variance is related to which one of the following?
(a) Homozygosity (b) Narrow sense heritability
(c) Transgressive segregation (d) All of the above

100. Selection is effective for those traits which are governed by one of the following?
(a) Additive genes (b) Dominant genes
(c) Epistatic genes (d) None of these

101. Dominance variance results due to which one of the following?
(a) Cytoplasmic effects (b) Inter allelic interaction
(c) Intra allelic interaction (d) None of these

102. Dominance variance has relationship with which one of the following?
(a) Heterosis (b) Heterozygosity
(c) Specific combining ability (d) None of these

103. Additive genetic variance results from which one of the following?
(a) Average effects of genes on all segregating
(b) Intra allelic gene interaction
(c) Inter allelic gene interaction
(d) None of these

104. Epistatic variance is fixable when the type of interaction is which one of the following?
(a) Additive x dominance (b) Additive x additive
(c) Dominance x dominance (d) None of these

105. Which one of the following scientist who discovered the role of chromosomes in sex determination?
(a) C. B. Bridges (b) Clarence Erwin McClung
(c) T. H. Morgan (d) Wilson and Stevens

106. Which one of the following is the genetic constitution of normal Drosophila female?
(a) XO (b) XX (c) XXY (d) XY

107. Sex chromosomes are also known as which one of the following?
(a) Allosomes (b) Autosomes (c) Lysosomes (d) Ribosomes

108. In Y chromosome of Melandrium, region II is known as which one of the following?
(a) Male fertility region (b) Male promoter region
(c) Male suppression region (d) Pairing region

109. Correns first reported cytoplasmic inheritance in which one of the following?
(a) Evening Prime Rose (b) Flour moth
(c) Four o'Clock plant (d) Water snail

110. The inheritance of characteristics coded for by genes found in the cytoplasm, typically from the mitochondria or chloroplasts is termed as Cytoplasmic inheritance. Cytoplasmic inheritance is also known as which one of the following?
(a) Extra chromosomal inheritance
(b) Extra nuclear inheritance
(c) Non-Mendelian inheritance (d) All of the above

111. Organeller inheritance and Uniparental inheritance are refers to which one of the following?
(a) Co-dominant inheritance (b) Cytoplasmic inheritance
(c) Intermediate inheritance (d) Sex-linked inheritance

112. Cytogenes are governed cytoplasmic inheritance, they are also known as which one of the following?
(a) Cytoplasmic genes (b) Extra nuclear genes
(c) Plasma genes (d) All of the above

113. Cytoplasmic genes are found in which one of the following?
(a) Chloroplasts (b) Mitochondria
(c) Both (a) and (b) (d) None of these

114. Which one of the following IOJAP (green and white stripes) leaf color is controlled in maize?
(a) Chloroplast genes (b) Mitochondrial genes
(c) Nuclear gene (d) None of these

115. Inheritance of sigma particle in Drosophila is an example of which one of the following?
(a) Cytoplasmic inheritance
(b) Inheritance due to ineffective particles
(c) Maternal effects (d) None of these

116. The molecule that encodes the genetic instructions in the all known living organisms is called as Deoxyribonucleic acid (DN(a). It is a polymer of which one of the following?
(a) Amino acids (b) Nucleosides
(c) Nucleotides (d) All of the above.

117. Thymine is always pair with which one of the following in a DNA molecule?
(a) Adenine (b)Cytosine
(c) Guanine (d) Uracil

118. The bases, adenine and thymine are joined by which one of the following in DNA molecule?
(a) Single hydrogen bonds (b) Double hydrogen bonds
(c) Triple hydrogen bonds (d) All of the above.

119. The bases, guanine and cytosine are joined by which one of the following in DNA molecule?
(a) Double hydrogen bonds (b) Triple hydrogen bonds
(c) Double phosphate bonds (d) Triple phosphate bonds

120. Purines are compounds with which one of the following type ring?
(a) Single ring (b) Double ring
(c) Triple ring (d) None of these

121. The uracil base is present in place of which one of the following base in RNA molecule?
(a) Adenine (b) Cytosine
(c) Guanine (d) Thymine

122. Which one of the following bases is not present in DNA and distinguishes with RNA?
(a) Adenosine (b) Cytosine
(c) Guanine (d) Uracil

123. Which one of the following bases is present in RNA that distinguishes to DNA?
(a) Adenosine (b) Cytosine
(c) Guanine (d) Uracil

124. RNA acts as a genetic material in which one of the following?
(a) E. coli (b) Neurospora
(c) TMV virus (d) None of these

125. Which one of the following RNA, acts as a carrier of amino acids during protein synthesis?
(a) mRNA (b) rRNA
(c) tRNA (d) None of these

126. Purines are nitrogen-containing ring compounds found in DNA and RNA includes which one of the following?
(a) Adenine and Cytosine (b) Adenine and Guanine
(c) Adenine and Thymine (d) Cytosine, Thymine and Uracil

127. Pyrimidines are nitrogen-containing ring compounds found in DNA and RNA includes which one of the following?
(a) Adenine and Cytosine (b) Adenine and Guanine
(c) Adenine and Thymine (d) Cytosine, Thymine and Uracil

128. The first man made cereal crop is Triticale, is an example of which one of the following?
(a) Artificial allopolyploidy (b) Artificial autopolyploidy
(c) Artificial evolution (d) Manmade crossing

129. The shortest of mitotic phases is which one of the following?
(a) Anaphase (b) Interphase
(c) Metaphase (d) Telophase

130. Which one of the following is the phenotypic ratio of a monohybrid cross in case of incomplete dominance?
(a) 1:1:2 (b) 1:2:1 (c) 1:3:1 (d) 3:1:1

131. Which one of the following haploid generation in an organism has a life cycle with alternation of generations?
(a) Gamete (b) Gametophyte
(c) Sporophyte (d) Zygote

132. At which one of the following stage, crossing over takes place during meiosis?
(a) Diakinesis stage (b) Diplotene stage
(c) Pachytene stage (d) Zygotene stage

133. A plasmid used to introduce BT toxin into a plant cell______.
(a) Is toxic for the plant cell
(b) Causes the plant to produce opines
(c) Carries a bacterial gene with a plant promoter
(d) Carries an antibiotic resistance gene that makes the plant resistant to antibiotics

134. When F1 plants are crossed with homozygous recessive plants, this is known as which one of the following?
(a) Criss-cross (b) Para-cross
(c) Recessive cross (d) Test cross

135. The homologous chromosomes follow the process of synopsis at which of following stage?
(a) Diplotene (b) Leptotene
(c) Pachytene (d) Zygotene

136. A transposable element is a specific DNA sequence transferred from one replicon to another. In plants, which one of the following transposable elements is not found?
(a) Ac (b) L1
(c) Tam3 (d) None of these

137. In meiotic cell division the daughter cells are produced by two successive division in which one of the following?
(a) First division is equational, second is reductional
(b) First division is reductional and second is equational
(c) Both divisions are equational
(d) Both divisions are reductional

138. Which one of the following is not support to Mendel's experiments using the garden pea?
(a) True-breeding varieties were scarce.
(b) They have relatively short generation time.
(c) Generative organs of the pea are enclosed within the flower.
(d) Could expect to observe segregation of traits among the offspring.

139. Pea plant belongs to one of the following?
(a) Self-pollination (b) Cross pollination
(c) Both (a) and (b) (d) None of these

140. One out of four plants had white flowers in Mendel's F2 generation because
(a) The trait is sex-linked
(b) One parent was homozygous recessive
(c) Both parents were heterozygous purple
(d) Both parents were heterozygous white

141. Which of the following is reason for difficulty to analyze Mendelian inheritance?
(a) Continuous variation (b) Epistasis
(c) Pleiotropic effects (d) All of the above

142. The 3:1 ratio in the F2 generation is really disguised by which one of the following?
(a) 1:1:2 (b) 1:2:1 (c) 2:1:1 (d) None of these

143. Which one of the following ratio is often referred as Mendelian ratio?
(a) 1:3:1 (b) 3:1 (c) 1:3:3:1 (d) None of these

144. Which one of the following is a dominant to recessive ratio in theF1 generation when Mendel crossed pure purple and white flowered pea plants?
(a) 3:1 (b) 4:0 (c) 4:0 (d) 4:1

145. Which one of the following dominant to recessive ratio in the in Mendel's F2 generation of the purple and white flower cross-ing?
(a) 1:1 (b) 1:3:1 (c) 3:1 (d) 9:3:3:1

146. Which one of the following is the name for a chromosome map unit?
(a) Centimorgan (b) Centistern (c) Centisutton (d) Decibarr

147. Testcross is represented by which of the following combina-tion?
(a) WW x WW (b) Ww x WW (c) ww x WW (d) Ww x Ww

148. Which one of the following is termed for an individual that has two of the same allele?
(a) Heteromologous (b) Heterozygous
(c) Homologous (d) Homozygous

149. Which one of the following character is referred by its appear-ance resulting from a combination?
(a) Alleleotype (b) Phenotype (c) Phototype (d) Stereotype

150. Which one of the following is unique characteristic of Mendel's work compared to previous geneticists?
(a) Choosing the garden pea for experiment
(b) Crossing true-breeding plants
(c) Keeping records of numbers of different progeny
(d) Producing hybrids

151. Which one of the following is modified Mendelian ratio?
(a) 1:3:1 (b) 3:1 (c) 9:3:3:1 (d) 9:7

152. The two alternatives for a trait are red and white and red is dominant. However, white is the most common trait. What is the genotype of a homozygous dominant individual?
(a) RR (b) rr (c) WW (d) ww

153. Which one of the following includes as early ideas about heredity
(a) Alleles do not influence each other
(b) Not all copies of a factor are identical
(c) Traits are transmitted directly
(d) Traits can be masked in some generations

154. In garden peas, height is determined by a single gene with tall being dominate to short. If two heterozygous plants are crossed, what proportion of the tall progeny will be homozygous dominant?
(a) 1/2 (b) 1/3 (c) 1/4 (d) 2/3

155. Signal codon refers to which one of the following codons?
(a) UAA and UAG (b) UAA, UAG and UGA
(c) UAA, UAG, UGA and AUG (d) None of these

156. Initiation of protein synthesis requires which one of the following?
(a) Activated amino acids attached to tRNA
(b) An energy source
(c) Initiation factor, ribosome sub units and mRNA
(d) All of the above

157. Both strands of the DNA molecule are used as template for the enzymatic formation of RNA is termed as which one of the following?
(a) Symmetrical transcription (b) Asymmetrical transcription
(c) Reverse transcription (d) None of these

158. Only one strand of the DNA molecule is used as template for the enzymatic formation of RNA is termed as which one of the following?
(a) Symmetrical transcription (b) Asymmetrical transcription
(c) Reverse transcription (d) None of these

159. Reverse transcription is also known as which one of the following term?
(a) Asymmetrical transcription (b) Symmetrical transcription
(c) Teminism (d) None of these

160. Reverse transcription has been reported in which one of the following?
(a) Certain tumour producing viruses (b) Drosophila
(c) E. coli (d) Neurospora

161. In 1902, gene action in man was first studied by which one of the following?
(a) Beadle and Ephrussi (b) Beadle and Tatum
(c) Brenner (d) Garrod

162. Abnormal pairing in codon occurs at the one of the following in wobble pairing?
(a) First base (b) Second base
(c) First and second bases (d) Third base

163. Which one of the following term is used for the process of synthesis of mRNA from DNA template?
(a) Transcription (b) Transduction
(c) Transformation (d) Translation

164. The process of DNA synthesis from RNA is termed as which one of the following?
(a) Reverse mutation (b) Reverse transcription
(c) Reverse translation (d) Transcription

165. The first case of mutation was discovered in which one of the following?
(a) Drosophila (b) Garden pea (c) Male lamb (d) Neurospora

166. Which one of the following refers to a unit of mutations in a gene?
(a) Cistron (b) Muton (c) Recon (d) None of these

167. Induced mutations are useful in which one of the following?
(a) Creation of variability
(b) Development of improved varieties
(c) Induction of male sterility
(d) Production of haploids

168. A mutation which kills 50% of the individuals that carry the mutation, it is referred to as which one of the following?
(a) Lethal mutation (b) Sub-lethal mutation
(c) Sub-vital mutation (d) Vital mutation

169. Which one of the following concerned with gene mutation?
(a) Gene, which causes another gene to undergo mutations.
(b) Mutant phenotype due to presence of mutant gene.
(c) Smallest unit of gene capable of undergoing mutation.
(d) All of the above

170. A gene which exhibits higher mutation rate than others is termed as which one of the following?
(a) Antimutator gene (b) Hot spot
(c) Mutable gene (d) Mutator gene

171. Visible autosomal mutants are detected in drosophila, by which one of the following?
(a) Attached X method (b) CIB method
(c) Curly lobe plum method (d) Muller 5 method

172. For induction of mutation, X-rays was used first by Stadler in which one of the following?
(a) Barley (b) Drosophila (c) Maize (d) Wheat

173. Substitution of one purine by another purine is termed as which one of the following?
(a) Addition (b) Deletion (c) Transition (d) Transversion

174. In tRNA molecule, three bases are complementary to the three bases of a specific codon in mRNA is termed as which one of the following?
(a) Anticodon (b) Codon (c) tRNA (d) None of these

ANSWERSHEET

3. Geneticvs

1. c	2. b	3. c	4. a	5. a	6. a	7. a
8. d	9. c	10. b	11. d	12. b	13. a	14. a
15. c	16. a	17. a	18. a	19. a	20. a	21. a
22. c	23. c	24. a	25. d	26. c	27. d	28. a
29. b	30. d	31. a	32. b	33. a	34. d	35. c
36. d	37. b	38. b	39. a	40. a	41. c	42. a
43. b	44. a	45. d	46. a	47. d	48. a	49. a
50. a	51. c	52. c	53. c	54. a	55. b	56. d
57. d	58. c	59. b	60. b	61. d	62. b	63. c
64. d	65. c	66. b	67. a	68. c	69. d	70. c
71. c	72. d	73. c	74. a	75. a	76. b	77. a
78. c	79. d	80. b	81. c	82. b	83. a	84. b
85. c	86. a	87. a	88. d	89. a	90. c	91. c
92. c	93. c	94. d	95. c	96. d	97. c	98. a
99. d	100. a	101. a	102. b	103. a	104. b	105. b
106. b	107. a	108. b	109. a	110. d	111. b	112. d
113. c	114. a	115. b	116. a	117. a	118. b	119. b
120. b	121. d	122. d	123. d	124. c	125. d	126. b
127. d	128. a	129. a	130. b	131. b	132. c	133. c
134. d	135. d	136. b	137. b	138. a	139. c	140. c
141. d	142. b	143. b	144. b	145. c	146. a	147. c
148. d	149. b	150. c	151. d	152. c	153. c	154. b
155. c	156. a	157. a	158. b	159. c	160. c	161 a
162. d	163. a	164. c	165. c	166. b	167. d	168. b
169. c	170. c	171. b	172. b	173. d	174. a	

4

Plant Breeding

1. Self-pollinated populations are generally which one of the following?
   (a) Heterozygous and Heterogeneous
   (b) Heterozygous and Homozygous
   (c) Homogeneous and Heterozygous
   (d) Homogeneous and Homozygous
2. Cross pollinated populations are generally which one of the following?
   (a) Heterogeneous and Heterozygous
   (b) Heterozygous and Homozygous
   (c) Homogeneous and Heterozygous
   (d) Homogeneous and Homozygous
3. Pure line selection is also known as which one of the following?
   (a) Crop production (b) Individual plant selection
   (c) Plant selection (d) Progeny selection
4. In the field of biological reproduction describing the fertilization of an ovum from one individual with the spermatozoa of another is called allogamy. Allogamy means crops are______.
   (a) Cross pollinated (b) Often cross pollinated
   (c) Self pollinated (d) No pollination
5. A method designed to concentrate favorable genes scattered among a number of individuals is termed as recurrent selection. The idea of recurrent selection developed who one of the following?
   (a) Hayes and Garber (b) Hull
   (c) Janssen (d) Jenkins
6. To develop which one of the following is, an objective of bulk, pedigree and back cross method of breeding?
   (a) Extraordinary variety (b) New variety
   (c) Pure line variety (d) Transform old to new variety

7. In mass selection, on their individual advantages, individual plants are selected and from the aggregate of their seeds, the next generation is propagated. Mass selection is based on which one of the following?
(a) Fraternal parent (b) Maternal parent
(c) Both (a) and (b) (d) None of these

8. Which one of the following considered by mass selection is/are?
(a) Usually applied to remove diseased plants, casual hybrids, or other defects.
(b) Easy method to select and multiply together individuals from a phenotypes mixture.
(c) Applied in cross-pollinating of vegetable species e.g. Carrots, Radishes, or Beetroots.
(d) All of the above

9. Which one of the following method developed by Hopkins in 1908, extensively used in maize?
(a) Bulk method (b) Ear to Row method
(c) Progeny test (d) Recurrent selection

10. Which one of the following considered with single seed descent (SS(d) methods?
(a) Artificial selection is based on the phenotype of individual plants, not on the progeny performance.
(b) Easy to maintain populations during inbreeding and natural selection cannot influence the population.
(c) Unless genotypes differ in their ability to produce viable seeds.
(d) All of the above

11. Mendel proposed which one of the following two laws?
(a) Law of inheritance & Law of genetics
(b) Law of segregation & Law of independent the assort ment of genes
(c) Law of segregation & Law of isolation
(d) None of these

12. Single seed descent (SS(d) method is a modification of which one of the following?
(a) Back cross method (b) Bulk method
(c) Mass selection (d) Pedigree method

13. In the method of breeding selection, which one of the following generation and beyond the pedigree is done?
(a) F2 (b) F3 (c) F4 (d) F5

14. Which one of the following refers to the grid method of selection?
(a) Mass selection (b) Pure mass selection
(c) Randomized mass selection (d) Stratified mass selection

15. Which one of the following is the simplest form of progeny selection?
(a) Back cross method (b) Ear to row method
(c) Mass selection (d) Recurrent selection

16. Which one of the following is a synonym, refers to the plant breeding?
(a) Crop Adjustment (b) Crop Enhancement
(c) Crop Improvement (d) Crop Technology

17. The process for killing the functional sexual parts of the plant (like pollen, ovule etc.) through using organic or inorganic chemicals is known as gametocides. These may be selective for male or female parts. Which one of the following is an application of gametocides in two-line method of rice breeding, except?
(a) Use of cytoplasmic male sterility (CMS)
(b) Use of environmentally induced genic male sterility (EGMS)
(c) Use of genic male sterility (GMS)
(d) Use of sporophytic and gametophytic male sterility (CGMS)

18. Some crops e.g., banana, do not produce true seed because of which one of the following?
(a) Male sterility (b) Female sterility
(c) Both (a) and (b) (d) None of these

19. Which one of the following is common and most useful in plant breeding as a technique for achieving cross-pollination?
(a) Male sterility (b) Female sterility
(c) Both (a) and (b) (d) None of these

20. In breeding, which one of the following considered with the types of heterosis is/are?
(a) F1 yielding more than the best yielding parents,
(b) F1 yielding more than the best yielding variety;
(c) F1 yielding more than the mean of the parents,
(d) All of the above

21. During 20th century, plant breeding has changed from selection based on the phenotype of individuals to selection based on the information derived at the ______________ level in molecular genetic laboratories and data from replicating field experiments.
(a) Deoxyribonucleic acid (DN(a) (b) Ribonucleic acid (DN(a)
(c) Both (a) and (b) (d) None of these

22. Plant breeding is often described as the art and science of developing which one of the following?
(a) Superior cultivars (b) Superior Seeds
(c) Superior Seeds (d) All of the above

23. Which one of the following is/are the fundamentals of maize breeding remain?
(a) Development of pure-lines by self-pollination.
(b) Production of crosses between derived lines.
(c) Production of the best hybrid for use by the farmer.
(d) All of the above

24. Which one of the following considered with maize breeding is/are?
(a) Germplasm improvement e.g. recurrent selection.
(b) Identification of hybrids having consistent.
(c) Reliable performance across an extensive number of env ronments.
(d) All of the above

25. The crop of hybrid is which one of following?
(a) Heterogeneous population of Heterozygous plants
(b) Heterogeneous population of Homozygous plants
(c) Homogeneous population of Heterozygous plants
(d) Homogeneous population of Homozygous plants

26. Which one of the following term is used for the transfer of pollen grains from anthers to stigma of a flowering plant?
(a) Chasmogamy (b) Male sterility
(c) Pollination (d) Self incompatibility

27. Five thousand pea plants grow as offspring from a series of identical dihybrid crosses. Which of the following most likely represent the data from an examination of all five thousand specimens, observing them for smooth/wrinkled peas (smooth is dominant) and for yellow/green peas (yellow is dominant)? The four phenotypes are listed in order in each case: smooth/yellow, smooth/green, wrinkled/yellow, wrinkled/green.
(a) 878, 2904, 886, 332 (b) 329, 2807, 928, 936
(c) 334, 937, 941, 2788 (d) 2798, 943, 935, 324

28. Most of the self-pollinating plants have small, relatively inconspicuous flowers that shed pollen directly onto the stigma, sometimes even before the bud opens. Generally, ancestors of self-pollinated crops are which one among the following?
(a) Cross pollinated crops (b) Often cross pollinated crops
(c) Both (a) and (b) (d) None of the above

29. Which one of the following is first dwarfing gene of wheat?
(a) Dee-Geo-Woo-Gen (b) Gregg 399
(c) Norin 10 (d) Tift 23A

30. Which one of the following considered with male sterility
(a) After opening of flower, fertilization as opposed to cleistogamy.
(b) Can be done in artificial culture i.e. In vitro.
(c) Controlled physiological hindrance to self-fertilization.
(d) Producing no functional pollen.

31. Who one of the following is known as soyabean breeder in India?
(a) E. E. Hartwig (b) K. Ramaiah
(c) T. S. Venkatraman (d) V. Santhanam

32. Which one of the following is an application of gametocides in three-line method of rice breeding?
(a) Use of cytoplasmic male sterility (CMS)
(b) Use of environmentally induced genic male sterility (EGMS)
(c) Use of genic male sterility (GMS)
(d) Use of sporophytic and gametophytic male sterility (CGMS)

33. Which one of the following regarding with pollination?
(a) After opening of flower, fertilization as opposed to cleistogamy.
(b) Can be done in artificial culture i.e. In vitro.
(c) Controlled physiological hindrance to self-fertilization.
(d) Producing no functional pollen.

34. Self incompatibility may consider with which one of the following?

(a) After opening of flower, fertilization as opposed to cleistogamy.

(b) Can be done in artificial culture i.e. In vitro.

(c) Controlled physiological hindrance to self-fertilization.

(d) Producing no functional pollen.

35. Which one of the following is true for chasmogamy?

(a) After opening of flower, fertilization as opposed to cleistogamy.

(b) Can be done in artificial culture i.e. In vitro.

(c) Controlled physiological hindrance to self-fertilization.

(d) Producing no functional pollen.

36. Inability to set seed from application of pollen produced on the same plant is termed as which one of the following?

(a) Chasmogamy (b) Pollination

(c) Self incompatibility (d) None of these

37. Biologist who believes that progeny develops from sperm only and sperm contribute to heredity is known as which one of the following?

(a) Animaleutists (b) Cytogenetist

(c) Ovists (d) None of these

38. Environmentally induced genic male sterility (EGMS) system is commonly used in rice. In EGMS systems, which one of the following is kinds of rice lines are made use, had been developed successfully in China?

(a) CMS (cytoplasmic male sterility)

(b) PGMS (Photosensitive Genic Male Sterility)

(c) TGMS (Thermosensitive Genic Male Sterility)

(d) Both (b) and (c)

39. The characters which express it suppressing the other characters are termed as which one of the following?

(a) Codominant (b) Dominant

(c) Partial dominant (d) None of these

40. Which one of the following is a barrier against inbreeding and the homozygosis caused by it?

(a) Self incompatibility (b) Male sterility

(c) Female sterility (d) Both (a) and (b)

41. With the addition of pollen may fail to germinate on the stigma, there are several mechanisms responsible for self-incompatibility in higher plants. Which one of the following is/are?
(a) A male gamete that enters the embryo sac may fail to unite with the egg cell.
(b) Pollen tube growth in the style may be inhibited to the extent that pollen fails to reach the ovary.
(c) Pollen tubes of sufficient length may fail to penetrate the ovule.
(d) All of the above

42. Which one of the following is an autopolyploid?
(a) Banana (b) Coffee
(c) Potato (d) All of the above

43. Who one of the following was an eminent scientists concerned with male sterility in plants?
(a) Johannsen (b) Jones and Davis
(c) Thomas Fairchild (d) None of these

44. In safflower, the cytoplasmic male sterility may not be used for because of which one of following?
(a) High cost of seed production (b) Low vigour in F1 plant
(c) Male Sterility in F1 plant (d) No heterosis in F1 plant

45. Without sexual fusion, development of embryo from egg cell is termed as which one of the following?
(a) Apogamy (b) Apospory
(c) Parthenogenesis (d) Unisexuality

46. Which one of the following is a type of apomixes, involve in the suppression of gametophyte formation resulting seeds are formed directly from somatic cells of the parent tissue?
(a) Apogamy (b) Apospory
(c) Both (a) and (b) (d) None of these

47. In some plants, development of a diploid embryo sac through integument cell without meiosis or somatic division of a nucellus. Which one of the following is a sort of agamospermy in which seed is produced without fertilization?
(a) Apogamy (b) Apospory
(c) Both (a) and (b) (d) None of these

48. From an egg development of an individual without fertilization, which one of the following occurs in only some plants?
(a) Apogamy (b) Apospory
(c) Parthenogenesis (d) None of these

49. Which one of the following is a type of gametophyte apomixis by which a diploid embryo sac cell result in a diploid embryo?
(a) Diploid parthenogenesis (b) Haploid parthenogenesis
(c) Matromorphy (d) Both (b) and (c)

50. Which one of the following is a condition, result of cytogenetic mechanisms occurring in the egg stage?
(a) Diploid parthenogenesis (b) Haploid parthenogenesis
(c) Both (a) and (b) (d) None of these

51. The unfertilized egg hatches and develops in which one of the following situation, normally to produce a viable male adult?
(a) Ameiotic parthenogenesis (b) Diploid parthenogenesis
(c) Haploid parthenogenesis (d) None of these

52. Meiosis has been entirely suppressed by which one of the following parthenogenesis?
(a) Ameiotic parthenogenesis (b) Haploid parthenogenesis
(c) Both (a) and (b) (d) None of these

53. A sequence of apomictic generations is followed by amphimictic generations in the life is termed as which one of the following parthenogenesis?
(a) Amniotic parthenogenesis (b) Cyclical parthenogenesis
(c) Diploid parthenogenesis (d) Haploid parthenogenesis

54. Hybrid varieties are produced by directing crosses. These varieties utilize which one of the following?
(a) GCA as component of heterosis
(b) SCA as component of heterosis
(c) Both (a) and (b) (d) None of the above

55. Which one of the following would be respective the chromosome number of the reproductive parts viz., endosperm, seed, embryo, pollen mother and megaspore mother cell?
(a) 2n, 3n, 2n, 2n (b) 2n, 3n, 3n, 3n
(c) 3n, 2n, 2n, 2n (d) 3n, 3n, 3n, 2n

56. By which one of the following, Vijpla, Vijay and Kalyan cotton varieties are developed?
(a) Back cross method (b) Bulk method
(c) Mass selection (d) Pedigree method

57. lsogenic lines are genetically identical (except for sex) purebreeding group of individuals, which are produced by which one of the following?
(a) Sib mating (b) Recurrent back crossing
(c) Random mating (d) Selfing

58. Linseed variety NP11 and NP12 are developed through which one of the following?
(a) Back cross method (b) Mass selection
(c) Pedigree method (d) Pure line selection

59. An operation involved in the production of hybrid varieties, which one of the following is concerned with hybrid varieties?
(a) Development of inbred lines (b) Evaluation of inbreeds
(c) Both (a) and (b) (d) None of these

60. Vasumati, Sumati, Chittimutyalu are recently released rice varieties through which one of the following breeding objectives?
(a) Aroma (b) Deep water
(c) Dryland conditions (d) Red rice

61. During hybrid seed production which one of following is considered, if the presence of male fertile plant in 'A' line with otherwise similar expression of all the distinguishing traits?
(a) B line plant (b) Pollen load
(c) Pollen shedder (d) R line plant

62. In self-pollinated crops, most commonly used method for handling of segregating generations is which one among the following?
(a) Bulk method (b) Mass selection
(c) Pedigree method (d) SSD method

63. Megaspores are a type of spore that present in heterosporous plants, also called macrospores. Megaspores are produced in ovules are known as which one of the following?
(a) Megagametogenesis (b) Megasporogenesis
(c) Microgametogenesis (d) Microsporogenesis

64. The seed of restorer line is produced in three line breeding by which one of following?
(a) AXB (b) AXR (c) BXR (d) RXR

65. Before carpels, the maturation of anthers is termed as protandry. Protandry is found in which one of the following crops?
(a) Maize and Sugarbeets
(b) Sorghum, Wheat and Cruciferous crops
(c) Sunflower and Carrot (d) Both (a) and (c)

66. Who was the scientist from the following first outlined in 1953, the method of developing multiline in wheat?
(a) Borlaug (b) Garber (c) Jenkins (d) Swaminathan

67. The theory of survival of fittest due to natural selection was proposed by which one of the following scientiast?
(a) August Weismann (b) Charles Darwin
(c) Gregor J. Mendel (d) None of these

68. In oat crops, the single seed descent (SS(d) method applied first in 1965 by which one of the following scientists?
(a) Garber (b) Graphius
(c) Vander plank (d) Vilmorin

69. Flowers open but only after pollination has taken place in which one of the following processes?
(a) Chasmogamy (b) Cleistogamy
(c) Dichogamy (d) Dicliny

70. Stigma receptive and pollen shed in closed flower like characteristics are of which one of the following?
(a) Chasmogamy (b) Cleistogamy
(c) Dichogamy (d) Dicliny

71. At different times, the maturation of male and female parts of a flower is termed as which one of the following?
(a) Chasmogamy (b) Cleistogamy
(c) Dichogamy (d) Dicliny

72. Before stamens phenomenon, maturation of pistils is termed as which one of the following?
(a) Protandry (b) Protocally
(c) Protogyny (d) Prototypy

73. Which one of the following scientists, SSD method was referred to as modified pedigree method?
(a) C. A. Brim (b) Goulden
(c) Graphius (d) Nilsson-Ehle

74. For intra-population improvement, except which one of the following is a method?
(a) Reciprocal full-sib selection
(b) Recurrent full-sib selection
(c) Recurrent half-sib selection
(d) Recurrent phenotypic selection

75. The testing of paired-plant crosses is done by which one of the following method of intra-population improvement?
(a) Recurrent full-sib selection
(b) Recurrent half-sib selection
(c) Recurrent mass selection
(d) Recurrent reciprocal selection

76. By use of their half-sib progeny, evaluation of individuals is considered with which one of the following method of intra-population improvement?
(a) Recurrent full-sib selection
(b) Recurrent half-sib selection
(c) Recurrent mass selection
(d) Recurrent reciprocal selection

77. By changing the frequency of polygenes, which one of the following is a method of breeding designed to increase the levels of quantitative variables (desirable qualities)?
(a) Recurrent full-sib selection
(b) Recurrent half-sib selection
(c) Recurrent mass selection
(d) Recurrent reciprocal selection

78. The vertifolia effect was discovered by Who one of the following in 1963, who named it after a potato cultivar of this name, in which the effect was very pronounced?
(a) Johannsen (b) Jones and Davis
(c) Thomas Fairchild (d) Van der Plank

79. Which one of the following is a recurrent selection breeding system, genetically different groups are maintained and individuals are mated from the different groups to test for combining ability in each selection cycle?
(a) Recurrent full-sib selection
(b) Recurrent half-sib selection
(c) Recurrent mass selection
(d) Recurrent reciprocal selection

80. The seeds from two individuals rather than one are used to form and for testing the new population. This is done only by which one of the following method of recurrent selection?
(a) Recurrent full-sib selection
(b) Recurrent half-sib selection
(c) Recurrent mass selection
(d) Recurrent reciprocal selection

81. Which one of the following is concerned with recurrent half-sib selection method?
(a) Evaluate the half-sib progeny from each plant.
(b) Inter-cross, elected individuals to form a new population.
(c) Both (a) and (b) (d) None of these

82. Which one of the following is a method of bypassing an incompatibility barrier between two genotypes or species by using a third genotype or species?

(a) Bridge cross (b) Composite cross
(c) Diallel cross (d) Both (a) and (b)

83. From the hybridization of several parents, a population derived either by the use of male sterility or by hand pollination is termed as which one of the following?

(a) Complete cross (b) Composite cross
(c) Compound cross (d) All of the above

84. Which one of the following is considered with the term compound cross?

(a) A combination of desirable genes from more than two inbred lines, breeding strains, or varieties.
(b) Crossing of an unidentified genotype with a phenotypically similar individual of defined genotype.
(c) Both (a) and (b) (d) None of these

85. Substitution of sexual reproduction by an asexual multiplication process without nucleus and cell fusion for seed production is known as

(a) Amphimixis (b) Apomixis
(c) Autogamy (d) Vegetative reproduction

86. Which one of the following terms is considered with crossing in all possible combinations of a series of genotypes?

(a) Complete cross (b) Compound cross
(c) Diallel cross (d) Both (a) and (c)

87. Crossing of an unidentified genotype with a phenotypically similar individual of defined genotype is termed as which one of the following?

(a) Bridge cross (b) Check cross
(c) Diallel cross (d) None of these

88. In which one of the following crops modified pedigree method of breeding was used?

(a) Rice (b) Sorghum
(c) Soybean (d) Wheat

89. In genetic analyses, backcross known as mattings through which a hybrid genotypes are used to determine linkage and crossing-over values. Which one of the following the back cross method cannot be used:?

(a) Development of varieties with multiple disease resistance.
(b) To enhance the yielding potential.
(c) Incorporation of exotic germplasm.
(d) Transfer of disease resistance.

90. Which one of the following is/are advantages of the pedigree method of breeding?

(a) If selection is effective, inferior genotypes may be discarded before lines are further evaluated.
(b) Selection in generation involves a different environment, which provides good genetic variability.
(c) Genetic relationship of lines is estimated and can be used to maximize genetic variability.
(d) All of the above

91. In pedigree method of breeding, which one of the following scientist introduced to the system of maintaining pedigree records?

(a) Borlaug (b) Harrington
(c) Hijalmer Nilsson (d) Vilmorin

92. Which one of the following person is related to "types of selection based on direction?"

(a) Brim (b) Graphius (c) Mather (d) Vilmorin

93. Which one of the following crop varieties have been developed by pedigree method?

(a) Cotton – Lakshmi, Digvijay,
(b) Sorghum – Co 18 and Tobacco – NP 222
(c) Wheat – NP-52 and Rice – Jaya, Padma (d) All of the above

94. With the help of mass pedigree method, crosses are made and further generations grown in bulk or as mass until suitable season occurs for making desirable selections against drought, insect and diseases etc. Which one of the following is/are merits of pedigree method?

(a) In plant's selection, an opportunity for the breeder to use skill and judgement.
(b) Take more time than bulk method to develop the new variety.
(c) Inferior plants and progenies are eliminated in early generations.
(d) Both (a) and (c)

95. Bulk method is also known as 'mass method' or 'population method of breeding.' in 1908, which one of the following scientists was first proposed to the bulk method?
(a) Harrington (b) Lonniquest
(c) Nilsson Ehle (d) Vilmorin

96. Molecular markers linked to quantitative trait loci (QTLs) have been used to investigate the relative significance of which one of the following is/are in heterosis?
(a) Dominance (b) Epistasis
(c) Overdominance (d) All of the above

97. For asexually propagated crops, which one of the following is: the best selection method?
(a) Back cross method (b) Clonal selection
(c) Mass selection (d) Pedigree method

98. Which one of the following is not concerned with mass selection?
(a) Adaptation of the original variety is not changed
(b) Does not retain considerable genetic variability in the new variety.
(c) Less demanding method
(d) Often extensive and prolonged yield trials are not necessary

99. Which one of the following term is used for 'seeds are formed but the embryos developed without fertilization?'
(a) Apomixis (b) Male sterility
(c) Self incompatibility (d) Self pollination

100. Grafting, gootee, budding and layering are examples of which one of the following?
(a) Bulbils (b) Sub-aerial stems
(c) Underground stems (d) Vegetative reproduction

101. A perfect/hermaphrodite flower has/have which one of the following?
(a) Pistil (b) Stamen
(c) Both (a) and (b) (d) None of these

102. First opening of a flower is termed as which one of the following?
(a) Anthesis (b) Xanthesis
(c) Zanthesis (d) None of these

103. For the improvement of which one of the following, pure line selection can be applied?
(a) Introduced varieties (b) Local varieties
(c) Old pure line varieties (d) All of the above

104. Rise to the endosperm is given by which one of the following?
(a) Sperm + Antipodal Cells (b) Sperm + Egg Cell
(c) Sperm + Secondary Nucleus (d) Sperm + Synergids

105. A large number of plants of similar phenotype are selected and their seeds are mixed together to constitute the new variety is termed as which one of the following selection procedure?
(a) Back cross (b) Bulk method
(c) Mass selection (d) Pedigree selection

106. A cross between two inbreeds by reversing the order of male and female parent is called which one of the following?
(a) Back cross (b) Reciprocal cross
(c) Test cross (d) None of these

107. A cross between a hybrid (F1 or a segregating generation) and one of its parents is known as which one of the following?
(a) Back cross (b) Mass cross
(c) Reciprocal cross (d) Test cross

108. Presence of male and female part on the same plant in the same flower is known as which one of the following?
(a) Monoecious (b) Hermaphrodite
(c) Dioecious (d) Male sterile

109. A cross of F1 with its homozygous dominant parent is known as which one of the following ratios?
(a) Reciprocal cross (b) Back cross
(c) Top cross (d) None of these

110. Relationship of parent-offspring is presented by which one of the following records?
(a) Parent record (b) Offspring record
(c) Pedigree record (d) Mass record

111. For segregating generations, which one of the following breeding methods is used?
(a) Back cross (b) Bulk method
(c) Pedigree method (d) All of the above

112. Genetically variable and homozygous populations of self-pollinated crops, as where one of the following breeding methods are applied?
(a) Mass selection (b) Pure-line selection
(c) Both (a) and (b) (d) None of these

113. An objective of a back cross method is to improve which one of the following?
(a) Adaptation of the variety
(b) Morphological characteristics
(c) One or two specific defects of a high yielding variety
(d) Yield

114. Recipient parent in the back cross method is also known as which one of the following?
(a) Donor parent (b) Non-donor parent
(c) Non-recurrent parent (d) Recurrent parent

115. Due to respect of which one of the following, mass selection and Pure-line selection are distinguished with each other?
(a) Genetic variation (b) New variety developed
(c) Progeny test performed (d) All of the above

116. For which one of the following, back cross method cannot be used?
(a) Inter-varietal transfer of simply inherited characters
(b) Production of NIL's
(c) Transgressive segregation (d) All of the above

117. Which one of the following is concerned with shuttle breeding?
(a) Between two or more locations, seeds to be grown at each.
(b) Instead of one, allows two breeding cycles per year.
(c) Both (a) and (b) (d) None of these

118. The concept of shuttle breeding is a proven contributor to the success of which one of the following?
(a) CIMMYT (b) ICAR (c) ICARDA (d) IRRI

119. Similar pure line, a blend of phenotypically that contributes different useful genes is termed as which one of the following?
(a) Madeline (b) Manyline
(c) Multiline (d) All of the above

120. The breeding strategy is termed as which one of the following that explain as "at two different locations, when breeding populations are grown i.e. they shuttle between the two locations?"
(a) Mass breeding (b) Mechanism breeding
(c) Shuttle breeding (d) Transfer breeding

121. Which one of the following persons were proposed to the evolutionary method of breeding?
(a) Gardner (b) Harlan
(c) Richey (d) Suneson

122. Handling of segregating generations for which one of the following type of crops, bulk method is useful?
(a) Cereals (b) Grain legumes and Oilseeds
(c) Smaller millets (d) All of the above

123. For which one of the following type of crop breeding, selected bulk method is mostly used?
(a) Bajra (b) Rice (c) Sorghum (d) Wheat

124. In the back cross method, character to be transferred must have which one of the following?
(a) Low heritability (b) Medium heritability
(c) High heritability (d) Both (a) and (b)

125. In the back cross method, donor parent is usually used as which one of the following?
(a) Once (b) Twice
(c) Thrice (d) None of these

126. The method is which one of the following for transfer of dominant and recessive gene in the case of backcross method?
(a) Different (b) Same
(c) Both (a) and (b) (d) None of these

127. Which one of the following is the source of dwarfing gene in rice?
(a) Dee-Geo-Woo-Gen (b) Gregg 399
(c) Norin-10 (d) Tift 23A

128. Colocasia antiquorum and Colocasia esculenta are an examples of which one of the following types?
(a) Corm (b) Tuber (c) Bulb (d) Rhizome

129. Chasmogamy is a botanical term describing a type of flower, which occurs in which one of the following types?
(a) Barley (b) Rice and Oats
(c) Wheat (d) All of the above

130. Which one of the following is considered with the term chasmogamy?
(a) Allows the flower to be cross-pollinated by another individual.
(b) Chasmogamy is the opposite of cleistogamy.
(c) Flower opens and exposes the stamens and styles.
(d) All of the above

131. Dicliny is the separation of male and female reproductive parts into different flowers. This condition occurs in which one of the following flowers?
(a) Pistillate (b) Staminate
(c) Both (a) and (b) (d) None of these

132. Which one of the following is considered with the term dicliny?
(a) Have female and male unisexual flowers on monoecy (same individuals).
(b) Have female and male unisexual flowers on dioecy (different individuals).
(c) Both (a) and (b)
(d) None of these

133. Who one of the following scientist regarded as the 'Father of Hybrid Rice' by developed the world's first the first hybrid rice varieties in the 1970?
(a) Norman E. Borlang (b) Yuan Longping
(c) Chandrakant T. Patel (d) T. S. Venkatraman

134. The aim of breeding method is to develop which one of the following types of varieties in case of self pollinated crops?
(a) Heterozygous variety (b) Homozygous variety
(c) Both (a) and (b) (d) None of these

135. The frequency of cross pollination varies greatly with which one of the following the factors?
(a) Location (b) Variety
(c) Weather conditions (d) All of the above

136. Due to which one of the following factors, pureline become genetically variable with time?
(a) Hybridization (b) Mechanical mixtures
(c) Mutation (d) All of the above

137. For the improvement of which one of the following types of crops, mass-pedigree selection method of breeding is particularly useful?
(a) Dioecious crops (b) Self incompatible crops
(c) Self pollinated crops (d) None of these

138. Before pollen is shed, removal of the anthers from a bud or flower is termed as emasculation. This is necessary for hybridization in which one of the following types of plants?
(a) Dioecious plants (b) Male sterile plants
(c) Self incompatible plants (d) Self pollinated plants

139. Which one of the following is considered with emasculation?
(a) Normal preliminary step in crossing to prevent self-pollination.
(b) Two basic methods i.e. individual and mass are used for emasculation.
(c) Both (a) and (b) (d) None of these

140. Mass emasculation is applied for which one of the following?
(a) In hemp by removing male plants from the dioecious population.
(b) In monoecious maize by mechanical detasseling.
(c) Both (a) and (b) (d) None of these

141. Mass emasculation is applied for which one of the following?
(a) In rice by spike treatment with high temperatures.
(b) In rice by spike treatment with warm water of about +42°C.
(c) Both (a) and (b) (d) None of these

142. Mass emasculation is applied for which one of the following?
(a) In tobacco by dark-phase treatment during anthesis.
(b) In wheat by using gametocides.
(c) Both (a) and (b) (d) None of these

143. Which one of the following varieties is/are an example of a heterozygous but homogenous population?
(a) Hybrid variety (b) Open pollinated variety
(c) Synthetic variety (d) All of the above

144. First artificial hybrid in plants was developed by which one of the following?
(a) Koelreutor (b) Thomas Fair child
(c) John knight (d) Naudin

145. With the genetic constitution AaBbccDd, a F1 plant is test crossed with aabbccdd. Which one of the following genotypes will be?
(a) 4 genotypes (b) 8 genotypes
(c) 10 genotypes (d) 12 genotypes

146. Who one of the following is an eminent scientists, transferred thick stem and high sugar contents from tropical noble cane to North Indian Canes. This process is known as noblization of sugarcane?
(a) C. T. Patel (b) Dharampal Singh
(c) Ram Dhan Singh (d) T. S. Venkatraman

147. Due to which one of the following, in wheat and rice the genetic purity of a variety can deteriorate?
(a) Delayed segregation (b) Inbreeding
(c) Mechanical contamination with seeds of other varieties
(d) Both (b) and (c)

148. A true breeding line obtained in cross pollinated species by continuous inbreeding is termed as which one of the following?
(a) Composite variety (b) Inbred
(c) Synthetic variety (d) Both (a) and (c)

149. By asexual reproduction, progeny of a single plant obtained is known as which one of the following?
(a) Clone (b) Inbred
(c) Pureline (d) Strain

150. The progeny of a single homozygous self pollinated plant is termed as which one of the following?
(a) Clone (b) Inbred
(c) Land race (d) Pureline

151. In an unopened bud, pollination and fertilization is the situation indicate to which one of the following terms?
(a) Chasmogamy (b) Cleistogamy
(c) Dichogamy (d) Homogamy

152. Which one of the following is correct regarding with chaemogamy?
(a) Stigma receptive and pollen shed after flower opens
(b) Stigma receptive and pollen shed in closed flower
(c) Both (a) and (b) (d) Neither (a) nor (b)

153. Which one of the following is true with cleistogamy?
(a) Stigma receptive and pollen shed after flower opens
(b) Stigma receptive and pollen shed in closed flower
(c) Both (a) and (b) (d) None of these

154. "Pollen shed before stigma receptive" is characteristics of which one of the following type flower?
(a) Protandry (b) Protgyny
(c) Thrum flower (d) None of these

155. "Stigma matures and ceases to be receptive before pollen shed" is characteristics of which one of the following type flower?
(a) Protandry (b) Protgyny
(c) Thrum flower (d) None of these

156. Development of seeds without sexual fusion is termed as which one of the following?
(a) Amphimixis (b) Apogamy
(c) Apomixes (d) Parthenogenesis

157. "Short styles and long stamens" is characteristics of which one of the following type flower?
(a) Thrum flower (b) Protandry
(c) Protgyny (d) None of these

158. Which of the following statements is correct regarding selfing?
(a) Decreases homozygosity
(b) Does not have any effect on heterozygosity
(c) Improves heterozygosity
(d) Reduces heterozygosity

159. Which one of the following scientists regarding epistasis hypothesis in 1952 suggested that influence of one locus on the expression of another may be involved in heterosis?
(a) Bateson (b) Gowen
(c) Johannsen (d) None of these

160. Due to which one of the following, cabbages are cross pollinated?
(a) Floral morphology (b) Protandry
(c) Protogyny (d) Self incompatibility

161. Due to which one of the following, cross pollination occurs in maize?
(a) Protandry (b) Protogyny
(c) Self incompatibility (d) None of these

162. Who one of the following scientist regarded as the 'Father of Green Revolution' in India?
(a) C. T. Patel (b) M. S. Swaminathan
(c) Norman E. Borlang (d) T. S. Venkatraman

163. For which one of the following, in backcross breeding repeated crossing with the recurrent parent is done?
(a) Break undesirable linkages.
(b) Completely recover the genotype of recurrent parent.
(c) Retain the intensity of desirable gene.
(d) Transfer the desirable gene from donor parent.

164. The progeny closely resembles of which one of the following in apomixes is?
(a) Male parent (b) Female parent
(c) Both male and female parent (d) None of these

165. By which one of the following pollination method, inbred lines in maize are maintained?
(a) Cross pollination (b) Open pollination
(c) Self pollination (d) Sib pollination

166. Which one of the following component lines are used in the development of composite variety of maize?
(a) Inbred lines (b) Open pollinated varieties
(c) Single crosses (d) Top crosses

167. Biologist who believes that progeny develops from the egg and sperm only serve as a stimulus is called as which one of the following?
(a) Animaleutists (b) Geneticists
(c) Ovists (d) None of these

168. A cross of F1 with its homozygous recessive parent is known as which one of the following?
(a) Reciprocal cross (b) Test cross
(c) Top cross (d) All of the above

169. For testing the general combining ability of an inbred line, which one of the following is the simplest method?
(a) Multiple cross (b) Single cross
(c) Test cross (d) Top cross

170. In Onion, which one of the following system is used for hybrid seed production?
(a) Cytoplasmic male sterility (b) Genetic male sterility
(c) Self-incompatibility (d) None of these

171. Based on the competitiveness of the genotypes under adverse environmental conditions, which one of the following breeding methods is?
(a) Bulk selection (b) Mass selection
(c) Pureline selection (d) Recurrent selection

172. A synthetic variety can be developed from, which one of the following?
(a) Inbreds (b) Clones
(c) Open pollinated varieties (d) All of the above

173. Which one of the following type of pollinations maintain to the synthetic varieties?
(a) Cross pollination (b) Open pollination
(c) Self pollination (d) Sib pollination

174. Which one of the following scientists were developed to the concept of disruptive mating and selection?
(a) Harrington (b) Mather and Thoday
(c) Vilmorin (d) None of these

175. The cultivar developed by combining selected cross pollinated lines is termed as which one of the following?
(a) Composite (b) Hybrid
(c) Inbred (d) Synthetic variety

176. Who one of the following scientists regarded as the 'Father of Green Revolution' in the world?
(a) C. T. Patel (b) M. S. Swaminathan
(c) Norman E. Borlang (d) T. S. Venkatraman

177. Formation of haploid archisporium cells by normal reductional division without pollination and formation of embryo sac without fertilization is termed as which one of the following terms?
(a) Apospory (b) Diplospory
(c) Parthenogenesis (d) Pseudogamy

178. A multiline variety is a mixture of which one of the following?
(a) Composites (b) Inbreds
(c) Several purelines (d) Synthetics

179. Weak stabilizing selection increases to which one of the following?
(a) Mean (b) Range
(c) Standard deviation (d) Variance

180. Selection is of how many types on basis of direction?
(a) 1 (b) 2 (c) 3 (d) 4

181. Which one of the following apomixis is an important in Allium, Fragaria, Agave?
(a) Non-recurrent (b) Recurrent
(c) Sporophytic (d) Vegetative

182. At one time, one of the two extremes, which one of the following selection favors?
(a) Directional selection (b) Disruptive selection
(c) Stabilizing selection (d) None of these

183. Stabilizing selection is also called as which one of the following?
(a) Formalizing selection (b) Normalizing selection
(c) Useful selection (d) Utilizing selection

184. In several species of Citrus, Garcinia, Euphorbia dulcis, Mangifera indica, which one of the following apomixis is an important?
(a) Non-recurrent (b) Recurrent
(c) Sporophytic (d) Vegetative

185. What is the meaning of 'Apo' in apomixes?
(a) Away (b) Far (c) Near (d) Nothing

186. The term 'Mixis' is used to define which one of the following?
(a) mingling (b) separation
(c) useless (d) None of these

187. In general gametophytic, which one of the following control to the apomixes?
(a) Dominant gene (b) Nuclear gene
(c) Recessive gene (d) None of these

188. Apomictic hybrids are called as which one of the following?
(a) Apohybrids (b) Sibhybrids
(c) Synhybrids (d) Vybrids

189. For combining ability, test is done by which one of the following?
(a) Open pollinated progeny test (b) Polycross test
(c) Top cross test (d) All of the above

190. GCA is tested in which one of the following?
(a) Composite (b) Synthetic
(c) Both (a) and (b) (d) None of these

191. Which one of the following is a quickest method of obtaining homozygous lines from a cross?
(a) Dihaploidy (b) Haploidy
(c) Trihaploidy (d) Xhaploidy

192. Tift 23A is which one of the following?
(a) Cytoplasmic genic male sterility in pearls millet.
(b) Genetic male sterility in sorghum cotton.
(c) New develop varieties of wheat. (d) None of these

193. Diploidization is achieved by treatment of which one of the following?
(a) Colchicine (b) Methionine
(c) Phenol (d) Urea

194. Which one of the following is a product in dihaploidy, SSD, Pedigree and bulk method of breeding?
(a) Cross pollinated species (b) Dihaploids
(c) Haploids (d) Pure breeding lines

195. Selection intensity is higher in which one of the following type of selection?

(a) Grid selection (b) Honey comb selection
(c) Both (a) and (b) (d) None of these

196. Yield is a character that concerned with which one of the following?

(a) High heritability (b) Low heritability
(c) Medium heritability (d) No heritability

197. In disruptive selection of the population is formed by individuals selected from which one of the following?

(a) Same class (b) Two classes
(c) More than one class (d) None of these

198. A selection, changes the frequency of alleles in a divergent manner is termed as which one of the following?

(a) Directional selection (b) Disruptive selection
(c) Stabilizing selection (d) None of these

199. A selection that removes individuals from a population is termed as which one of the following?

(a) Directional selection (b) Disruptive selection
(c) Stabilizing selection (d) None of these

200. Which one of the following term is used for the selection resulting in a shift in the population mean in the direction desired by the breeder?

(a) Directional selection (b) Disruptive selection
(c) Stabilizing selection (d) None of these

201. Which selection favours more than one phenotypic optimum (and different genotypes) and discriminates against intermediate is called as which one of the following?

(a) Directional selection (b) Disruptive selection
(c) Stabilizing selection (d) None of these

202. Which one of the following is/are considered regarding the disruptive selection?

(a) From both ends of a phenotypic distribution divided by deviation of a sample of means.
(b) Leading to the fixation of alternative alleles in members of the population.
(c) Result after several generations of selection, two divergent phenotypic extremes within population.
(d) Both (b) and (c)

203. Which one of the following is/are considered regarding the stabilizing selection?
(a) From both ends of a phenotypic distribution divided by deviation of a sample of means.
(b) Leading to the fixation of alternative alleles in members of the population.
(c) Result after several generations of selection, two divergent phenotypic extremes within population.
(d) None of these

204. For improvement of a population, when two or more than two methods are used in the selection programme is called as which one of the following?
(a) Combined selection (b) Nil Selection
(c) Thrice selection (d) Twin selection

205. The inter-population scheme is called as which one of the following?
(a) Reciprocal recurrent selection
(b) Recurrent full sib selection
(c) Recurrent half sib selection (d) None of these

206. Which one of the following is improved through reciprocal recurrent selection?
(a) One population (b) Two population simultaneously
(c) Three populations (d) Four populations randomly

207. Vegetative reproduction is also referred to as which one of the following?
(a) Clonal propagation (b) Leaf propagation
(c) Root propagation (d) Stem propagation

208. Which one of the following is an example of protogyny?
(a) Maize (b) Oryza sativa (c) Plantago spp. (d) Wheat

209. The stamens and pistils are of unequal length in floral modification is termed as which one of the following?
(a) Distyly (b) Heterostyly
(c) Sinstyly (d) Unstyly

210. Which one of the following is concerned with the term heterostyly?
(a) A polymorphism among flowers that ensures cross-fertilization through pollination by visiting insects.
(b) At a single locus, pin ? pin and thrum ? thrum crosses are incompatible due to alleles.
(c) Flowers have anthers and styles of different length.
(d) Both (a) and (c)

211. Which one of the following is concerned with the term distyly?
(a) A polymorphism among flowers that ensures cross-fertilization through pollination by visiting insects.
(b) Flowers have anthers and styles of different length.
(c) At a single locus, pin ? pin and thrum ? thrum crosses are incompatible due to alleles.
(d) Both (b) and (c)

212. Distyly refers to the presence of either pin or thrum flowers. Morphology is generally controlled by alleles, which one of the following is correct for the term distyly?
(a) Pin morphology is controlled by a recessive alleles s
(b) Thrum morphology is controlled by a dominant alleles S
(c) Both (a) and (b) (d) None of these

213. Who one of the following scientist regarded as the 'Sugarcane Breeder?'
(a) Norman E. Borlang (b) M. S. Swaminathan
(c) C. T. Patel (d) T. S. Venkatraman

214. Pin flowers have which one of the following?
(a) Long pistils and no stamens
(b) Long pistils and short stamens
(c) No pistils and short stamens
(d) Short pistils and long stamens

215. For economically useful crops, the desired improved varieties of are raised by which one of the following?
(a) Hybridization (b) Mutation
(c) Natural selection (d) All of the above

216. To exploit hybrid vigour in high yielding hybrid crop varieties, farmers have need to purchase fresh hybrid seed every year due to which one of the following reasons?
(a) It is cheaper to purchase fresh seed.
(b) The Government of India has accepted Dunkel's proposal.
(c) The hybrid vigour is lost due to inbreeding depression.
(d) They are not allowed to grow their own seeds.

217. Pangenesis is a hypothetical mechanism for heredity. Which one of the following scientist coined this term in 1868?
(a) Charles Darwin (b) Thomas Hunt Morgan
(c) Wilhelm Hofmeister (d) Wilhelm Johannsen

218. Changing the genetics of plants in order to produce desired characteristics is known as plant breeding. The main aims of plant breeding are to produce which one of the following?
(a) Disease free varieties (b) Early maturing varieties
(c) High yielding varieties (d) All of the above

219. Genetic modification of plants is achieved by which one of the following to produce a desirable phenotype?
(a) Adding a specific gene or genes to a plant
(b) By knocking down a gene with RNAi
(c) Both (a) and (b) (d) None of these

220. Modern plant breeding is using many techniques of molecular biology, results is/are which one of the following?
(a) Increased quality and yield of the crop.
(b) To insert, desirable traits into plants.
(c) Both (a) and (b) (d) None of these

221. Scientists are trying to get hybridization between tomato and potato. Which one of the following name most accurate would be?
(a) Mopato (b) Pomato
(c) Tomepo (d) Topemo

222. Which one of the following is the oldest breeding method?
(a) Hybridization (b) Introduction
(c) Mutation breeding (d) Selection

223. The reason why vegetatively reproducing crop plants are best suited for maintaining Hybrid vigour is that
(a) Once a desired hybrid, has been produced there are few chances of losing it.
(b) They are more resistance to diseases
(c) They can be easily propagated
(d) They have a longer life span

224. By which one of the following improved varieties of wheat have been developed suitable for Indian climates?
(a) Cloning of polyploids
(b) Hybridization and mutation
(c) Mutation and cloning
(d) Polyploidy and hybridization

225. The emasculation of flower buds is achieved by the removal of which one of the following?
(a) Anthers (b) Corolla
(c) Sepals (d) Stamens

226. A novel technique devised to produce vast quantities of strong and healthy plantlets by rapid vegetative multiplication under controlled conditions.
(a) Anther culture (b) Micro propagation
(c) Polymerase chain reaction (d) Protoplast fusion

227. Which one of the following term is used for any reproductive process that requires pollination but does not involve male inheritance?
(a) Allogamy (b) Autogamy
(c) Pseudogamy (d) Syngamy

228. Which one of the following influence to the anthesis?
(a) Ambient temperature
(b) Day length
(c) Duration and intensity of illumination
(d) All of the above

229. Which one of the following is an example of Cleistogamy?
(a) Oxalis (b) Viola
(c) Wheat (d) All of the above

230. Among the forage grasses, berseem (Trifolium alexandrinum) is classified under which one of the following categories?
(a) Cross pollinated (b) Self-incompatible
(c) Self-pollinated (d) Male sterile

231. Which one of the following people proposed to the honey comb design for selection within crosses?
(a) Fasoulas (b) Harrington
(c) Mather (d) Shull

232. Which one of the following technology that allow the specific identification, isolation, and alteration of genes and their reintroduction into living organisms to produce transgenic varieties?
(a) Genetic engineering (b) Plant tissue culture
(c) Recombinant DNA techniques (d) None of these

233. Who one of the following is known as forage breeder?
(a) G.W. Burton (b) N. G. P. Rao
(c) T. J. Jenkin (d) Both (a) and (c)

234. Which one of the following is considered with Flavr Savr tomato?

(a) Flavr Savr tomato is also known as CGN-89564-2.

(b) Cultivars using an antisense polygalacturonase gene isolate from tomato.

(c) Flavr Savr tomato was developed and commercialized on May 18, 1994.

(d) All of the above

235. Which one of the following was developed through the use of antisense RNA to regulate the expression of the enzyme polygalacturonase (PG) in ripening fruit?

(a) CGN-89564-2 (b) Flavr Savr tomato

(c) Pomato (d) Both (a) and (b)

236. Which of the following is true about evolution theory?

(a) Darwin's theory of evolution was through natural selection.

(b) Evolution is an ability of population to undergo evolutionary divergence.

(c) Evolution leads to changes in genetic diversity.

(d) All of the above

237. For self-pollinated crops, which one of the following is intra-population improvement?

(a) Back cross selection (b) Mass Selection

(c) Pure Lines Selection (d) Both (b) and (c)

238. For cross-pollinated crops, except which one of the following is intra-population improvement?

(a) Composite Varieties (b) Family Selection

(c) Mass Selection (d) S1 Selection

239. Except which one of the following is an inter-population improvement for self-pollinated crops?

(a) Bulk selection (b) Pedigree selection

(c) Pure Lines Selection (d) Single seed decent

240. Except which one of the following is an inter-population improvement for cross-pollinated crops?

(a) Hybrid Varieties (b) Modified Mass Selection

(c) Rec.Rec.Selection (d) Synthetic Varieties

241. Which one of the following are types of hybrid verities?

(a) Single cross (IA x I(b)

(b) Double cross (IA x I(b) x (IC x I(d)

(c) Three way cross (IA x I(b) x IC

(d) All of the above

242. The cytoplasmic male sterility (CMS) gene used in modern wheat breeding is derived from which one of the following wild tetraploid variety?

(a) Triticum monococcum, (b) Triticum tauschii
(c) Triticum timopheevii (d) None of these

243. Match the following Indian Plant Breeders with the suitable match as they are regarded.

1. T. S. Venkatraman A. Potato breeder
2. B. P. Pal B. Cotton breeder
3. Pushkarnath C. Sugarcane breeder
4. K. Ramaiah D. Wheat breeder
5. V. Santhanam E. Rice breeder

(a) 1B 2C 3D 4A 5E (c) 1C 2D 3A 4E 5B
(b) 1C 2D 3B 4A 5E (d) 1D 2C 3A 4B 5E

244. Match the following Indian Plant Breeders with the suitable match as they are regarded.

1. N. G. P. Rao A. Sorghum breeder
2. Ram Dhan Singh B. Wheat breeder
3. D. S. Athwal C. Pearl millet breeder
4. C. T. Patel D. Cotton breeder
5. Dharampal Singh E. Oil-seed breeder

(a) 1B 2C 3D 4A 5E (c) 1C 2D 3A 4E 5B
(b) 1C 2D 3B 4A 5E (d) 1D 2C 3A 4B 5E

245. Match the following apomixes with the suitable descriptions.

1. Non-recurrent A. Embryos do not arise from the cells of the gametophyte; they arise from cells of nucellus or the integument.
2. Recurrent B. The flowers are replaced by bulbils which frequently germinate while still on the plant
3. Sporophytic C. Arise either from the egg (haploid parthenogenesis) or from some other cell of the gametophyte (haploid apogamy).
4. Vegetative D. Generally arises either from an archesporial cell or from some other part of the nucellus.

(a) 1B 2C 3D 4A (c) 1C 2D 3A 4B
(b) 1C 2D 3B 4A (d) 1D 2C 3A 4B

246. Match the following polyploid types with their suitable examples.

1. Triploid — A. Dahlias (Dahlia hortensis)
2. Tetraploid — B. Kiwifruit (Actinidia delicios(a)
3. Hexaploid — C. Seedless watermelons (Citrullus lanatus)
4. Octaploid — D. Cotton (Gossypium hirsutum)
5. Decaploid — E. Strawberries (Fragaria ananass(a)

(a) 1B 2C 3D 4A 5E (c) 1C 2D 3A 4E 5B
(b 1C 2D 3B 4A 5E (d) 1D 2C 3A 4B 5E

247. Match the following polyploid types with their suitable examples of crops.

1. Triploid — A. Plumed cockscomb (Celosia argente(a), Cordgrass (Spartina anglic(a)
2. Tetraploid — B. Banana, citrus, ginger
3. Hexaploid — C. Potato, cabbage, leek, tobacco, peanut, kinnow
4. Octaploid — D. Chrysanthemum, bread wheat, triti cale, oat
5. Dodecaploid (n=12) — E. Sugarcane, oca (Oxalis tuberos(a), pansies (Viola wittrockian(a)

(a) 1B 2C 3D 4E 5A (c) 1C 2D 3A 4E 5B
(b) 1C 2D 3B 4A 5E (d) 1D 2C 3A 4B 5E

*Apples (Triploi(d) and Kinnows (Tetraploi(d) can be diploid, triploid, or tetraploid.

248. Match the following name of the scientists with their contributions.

1. Allard and Bradshaw — A. Recurrent Selection for GCA
2. Hull — B. First artificial hybrid
3. Jenkins — C. G (Genotypes) x E (Environ ments) interaction
4. Shull — D. Recurrent Selection for SCA
5. Thomas Fairchild — E. Heterosis

(a) 1B 2C 3D 4A 5E (c) 1C 2D 3A 4E 5B
(b) 1C 2D 3B 4A 5E (d) 1D 2C 3A 4B 5E

249. Match the following contributions with the name of the scientists.

1. Pureline concept — A. Jones
2. Double cross scheme — B. Vilmorin
3. Backcross method — C. Harlan and Pope
4. Progeny test — D. Johannsen
5. Ear to Row method — E. Hopkins

(a) 1B 2C 3D 4A 5E (c) 1C 2D 3A 4E 5B
(b) 1C 2D 3B 4A 5E (d) 1D 2C 3A 4B 5E

250. Match the following eminent scientists with their scientific contributions.

1. Jones and Davis — A. Components of genetic variance classification
2. Fischer — B. Sprophytic System of self-in compatibility
3. Lewis — C. Centres of diversity, Law of homologus series
4. Vavilov — D. Self-incompatibility, Self-incom patibility classification
5. Hughes and Babcock — E. Cytoplasmic Genetic Male ste rility

(a) 1E 2A 3D 4C 5B (c) 1B 2E 3A 4D 5C
(b) 1E 2A 3B 4D 5C (d) 1E 2B 3A 4C 5D

251. Match the following eminent scientists with their scientific contributions.

1. Davenport — A. Raphano brassica
2. Flor — B. Dominance hypothesis
3. Karpenchenko — C. Modified bulk method
4. Nilsson and Ehle — D. Bulk method
5. Atkins — E. Gene for gene hypothesis

(a) 1E 2A 3D 4C 5B (c) 1B 2E 3A 4D 5C
(b) 1E 2A 3B 4D 5C (d) 1E 2B 3A 4C 5D

252. Match the following scientific contributions with their eminent scientists.

1. Chemical mutagen — A. Jenson
2. Multiline concept — B. Blackslee and Nebel
3. Colchicine — C. Yuan Long Ping
4. Mutation — D. Hugo de Vries
5. Father of hybrid rice — E. Aurbach

(a) 1E 2A 3D 4C 5B (c) 1B 2E 3A 4D 5C
(b) 1E 2A 3B 4D 5C (d) 1E 2B 3A 4C 5D

253. Match the following scientists with their eminent scientific contributions.

1. C. T. Patel — A. Male sterility in maize
2. Painter — B. Mechanism of insect resistance
3. Rhoades — C. Single Seed Descent Method
4. Goulden — D. Semi dwarf rice varieties at IRRI
5. T.T. Chang — E. World's first cotton hybrid in 1970

(a) 1E 2A 3D 4C 5B
(b) 1E 2A 3B 4D 5C
(c) 1B 2E 3A 4D 5C
(d) 1E 2B 3A 4C 5D

254. Match the following genetic populations with their brief description in plant breeding.

1. Homogeneous — A. Genetically similar population
2.Heterogeneous — B. Populations segregate on selfing
3. Homozygous — C. Genetically dissimilar population
4. Heterozygous — D. Non-segregating populations

(a) 1A 2C 3D 4B
(b) 1A 2D 3B 4C
(c) 1D 2B 3A 4C
(d) 1B 2C 3A 4D

255. Match the following combinations of genetic populations with their brief description in plant breeding.

1. Homogeneous and Homozygous — A. Genetically dissimilar but non-segregating populations
2. Homogeneous and Heterozygous — B. Genetically similar but segre gating on selfing
3. Heterogeneous and Homozygous — C. Genetically dissimilar and seg regating populations
4. Heterogeneous and Heterozyous — D. Genetically similar and non-segregating populations

(a) 1A 2C 3D 4B
(b) 1A 2D 3B 4C
(c) 1D 2B 3A 4C
(d) 1B 2C 3A 4D

256. Match the following combinations of genetic populations with their brief description in plant breeding.

1. Homogeneous and Homozygous — A. Pure lines and inbred lines
2. Homogeneous and Heterozygous — B. Multi lines and mass selected varieties in autogamous species
3. Heterogeneous and Homozygous — C. Composites and synthetics
4. Heterogeneous and Heterozyous — D. F1 hybrids between inbred lines and progeny of a clone

(a) 1A 2C 3D 4B
(b) 1A 2D 3B 4C
(c) 1D 2B 3A 4C
(d) 1B 2C 3A 4D

257. Match the following recently released rice varieties with their breeding objectives.

1. Swarna, Vijetha, Jaya	A. Resistant to blast diseases
2. Vikas, Swarnamukhi	B. Resistant to salinity
3. Somasila Swathi, Shravani	C. Resistant to pests BPH
4. Vajram, Vijetha Deepti, Krishnaveni	D. Resistant to pests gall midge
5. Kavya, Surekha Jagtial sannalu	E. High Yielding Vari eties

(a) 1E 2B 3A 4C 5D
(b) 1A 2E 3B 4C 5D
(c) 1A 2E 3D 4B 5C
(d) 1B 2E 3A 4D 5C

258. Match the following Lycopersicon spp. with their characteristics.

1. L. pimpinellifolium	A. Fusarium wilt, early blight re sistant
2. L. peruvianum	B. Fruit borer resistant
3. L. cheesmanii	C. Drought tolerant
4. L. hirsutum	D. Salt resistant
5. L. pennellii	E. Leaf curl virus resistant

(a) 1A 2E 3D 4B 5C
(b) 1A 2E 3B 4C 5D
(c) 1E 2B 3A 4C 5D
(d) 1B 2E 3A 4D 5C

ANSWERSHEET

4. Plant Breeding

1. d	2. a	3. b	4. a	5. a	6. c	7. b
8. d	9. b	10. b	11. d	12. b	13. a	14. d
15. b	16. d	17. b	18. a	19. d	20. a	21. d
22. d	23. d	24. c	25. c	26. d	27. a	28. c
29. d	30. a	31. d	32. b	33. c	34. a	35. c
36. a	37. d	38. b	39. a	40. d	41. b	42. c
43. c	44. a	45. b	46. c	47. d	48. a	49. c
50. a	51. c	52. a	53. b	54. c	55. c	56. a
57. b	58. d	59. c	60. a	61. c	62. c	63. b
64. d	65. d	66. a	67. b	68. b	69. a	70. b
71. c	72. c	73. a	74. a	75. a	76. b	77. c
78. d	79. d	80. a	81. c	82. a	83. b	84. a
85. b	86. d	87. b	88. c	89. b	90. d	91. c
92. c	93. d	94. d	95. c	96. b	97. b	98. b
99. a	100. d	101. c	102. a	103. d	104. c	105. c
106. b	107. a	108. b	109. b	110. c	111. c	112. d
113. c	114. d	115. d	116. d	117. c	118. a	119. c
120. d	121. d	122. d	123. d	124. c	125. a	126. b
127. a	128. a	129. d	130. d	131. c	132. c	133. b
134. b	135. d	136. d	137. c	138. d	139. c	140. c
141. c	142. c	143. a	144. b	145. b	146. d	147. c
148. b	149. a	150. d	151. b	152. a	153. b	154. a
155. b	156. c	157. a	158. d	159. b	160. d	161 a
162. b	163. b	164. b	165. c	166. b	167. c	168. b

169. d	170. a	171. a	172. d	173. b	174. b	175. b
176. c	177. c	178. c	179. d	180. c	181. d	182. a
183. b	184. c	185. a	186. a	187. a	188. d	189. d
190. b	191. a	192. a	193. a	194. d	195. a	196. b
197. c	198. b	199. c	200. a	201. b	202. d	203. a
204. a	205. a	206. b	207. a	208. c	209. b	210. d
211. b	212. c	213. d	214. b	215. a	216. c	217. a
218. d	219. c	220. c	221. b	222. b	223. a	224. b
225. a	226. b	227. c	228. d	229. d	230. c	231. a
232. c	233. d	234. d	235. d	236. d	237. d	238. a
239. c	240. b	241. d	242. c	243. c	244. c	245. c
246. b	247. a	248. c	249. d	250. a	251. c	252. b
253. d	254. a	255. c	256. b	257. a	258. a	

5

Plant Classification

1. The plant classification is deal by which one of the following sciences?
(a) Anatomy (b) Palaeobotany
(c) Taxonometrics (d) Taxonomy

2. Which one of the following system of classification is related with Bentham and Hooker's system?
(a) Artificial (b) Genetic
(c) Natural (d) Phylogenetic

3. Who is known as father of Modern taxonomy?
(a) Carolus Linnaeus (b) Friedrich Alefeld
(c) M. S. Swaminathan (d) Philip Miller

4. Who one of the following people is the regarded father of binomial classification?
(a) Bentham (b) Hooker
(c) Hutchinson (d) Linnaeus

5. The plants from the Mediterranean flora with medicinal description were added in ancient literate period by which one of the following?
(a) Albertus Magnus (b) Dioscorides
(c) Hooker (d) John Gerard

6. Who written the "Families of Flowering Plants"
(a) Bentham and Hooker (b) Bessey
(c) Engler (d) Hutchinson

7. Who first coined the word taxonomy?
(a) de Candolle (b) Hutchinson
(c) John Ray (d) Linnaeus

8. The mathematical methods helps to classify the plants is known as which one of the following branch of science?
(a) Biometrics (b) Quantitative taxonomy
(c) Systematics (d) Taxonometrics

9. When plants were classified on the basis of usefulness in the history of taxonomy, this period is known as which one of the following?
(a) Ancient literate (b) Medieval
(c) Preliterate (d) Theory of evolution

10. Who one of the following people is the regard as the father of botany?
(a) Albertus Magnus (b) Dioscorides
(c) Linnaeus (d) Theophrastus

11. To systematize a local flora by which one of the following first documented efforts?
(a) Bentham (b) John Gerard
(c) Linnaeus (d) Theophrastus

12. Which one of the following written the works "Enquiry into Plant" and "The Causes of Plants"
(a) Albertus Magnus (b) Dioscorides
(c) Pliny the Elder (d) Theophrastus

13. A multivolume work on the name of Historia Naturalis was written by
(a) Albert Magnus (b) Linnaeus
(c) Pliny the Elder (d) Theophrastus

14. Which one of the following written to Materia Medica?
(a) Albert Magnus (b) Dioscorides
(c) Hennyman (d) Pliny the Elder

15. For the first time, monocot and dicot the classification system was recognized by which one of the following persons?
(a) Albert Magnus (b) Bentham and Hooker
(c) Dioscorides (d) Linnaeus

16. The first description produced for which one of the following by John Gerard (1542-1612)
(a) Onion (b) Potato
(c) Sweet potato (d) Tomato

17. 'Sytema Naturae' written by C. Linnaeus contained a list of which one of the following?
(a) 4200 species of animals (b) 4200 species of plants
(c) 5000 species of plants (d) 8000 species of plants

18. The classification and nomenclature of the plants are deals with which one of the following science?
(a) Anatomy (b) Paleobotany
(c) Systematic Botany (d) Taxonomy

19. Taxonomic group of any rank or unit is known as which one of the following?
(a) Fauna (b) Flora
(c) Plantae (d) Taxon

20. All the important characters of the plant and plants are classified according to their affinities; the system of classification is called as which one of the following?
(a) Artificial (b) Morphological system
(c) Natural system (d) Phylogenetic system

21. According to their evolutionary and genetic affinities, plants are classified. Classification of this system is known as which one of the following?
(a) Evolutionary system (b) Genetic system
(c) Natural system (d) Phylogenetic system

22. The classification system in which only few characters of the plants are being considered is called as which one of the following system?
(a) Artificial system (b) Morphological system
(c) Natural system (d) Phylogenetic system

23. First time plants were grouped into perfecti and imperfecti groups on the basis of presence or absence of flowers by which one of the following person?
(a) Jean Bauhin (b) Jerome Bock
(c) Otto Brunfels (d) Pliny

24. Which one of the following people classified to the plants as trees, shrubs and herbs?
(a) Jerome Bock (b) John Ray
(c) Otto Brunfels (d) William Turner

25. "Flower clock" was a garden plan hypothesized by of which one of the following scientist?
(a) Carolus Linnaeus (b) Mendal
(c) Philip Miller (d) Whitteker

26. Who one of the following persons is regarded Father of English Botany?
(a) Bentham (b) Jerome Bock
(c) Linnaeus (d) William Turner

27. About the British herbals "Herbal or General Historie of Plante" was written by which one of the following persons?
(a) Hooker (b) Jean Bauhin
(c) John Gerard (d) John Ray

28. "Historia Plantarum Universalis" contained the descriptions of about 5000 plants written by which one of the following persons?
(a) G. Bauhin (b) Jean Bauhin
(c) John Gerard (d) John Ray

29. Which one of the following persons is Swiss physician written to "An Introduction to Botany" in which 6000 plant species were described?
(a) G. Bauhin (b) Jean Bauhin
(c) John Gerard (d) John Ray

30. On the basis of the characters of the seed and embryo "De Plantis" contains classification of 1500 plants was written by which one of the following persons?
(a) Andrea Cesalpino (b) G. Bauhin
(c) John Ray (d) Kobert Morison

31. In "Plantarum Historia Universalis" woody and herbaceous plants were differentiated by which one of the following persons?
(a) Jean Bauhin (b) John Gerard
(c) John Ray (d) Kobert Morison

32. In 14th century, after the medieval age or Dark age the works on taxonomy again started. This period is known as which one of the following periods?
(a) Modern (b) Renaissance
(c) Systematic (d) Theory of Evolution

33. In Kew Royal Botanical Garden, England, important system of classification was established by which one of the following persons?
(a) Bentham and Hooker (b) de Candolle
(c) Jean Bauhin (d) Linnaeus

34. Who one of the following persons is father of modern botany?
(a) Bentham (b) Darwin
(c) Hooker (d) Linnaeus

35. Which one of the following persons is concerned with the binomial system of classification?
(a) Adanson (b) Bentham
(c) Candolle (d) Linnaeus

36. An artificial sexual system of classification proposed by Linnaeus, contains how many classes?
(a) 16 classes (b) 20 classes (c) 24 classes (d) 30 classes

37. In which one of the following Linnaeus's systems of classification published?
(a) An Introduction to Botany (b) Genera Plantarum
(c) Historia Plantarum (d) Plantarum Historia

38. Modern system of classification started from the work of which concerned persons?
(a) Bentham and Hooker (b) Candolle
(c) Hutchinson (d) Linnaeus

39. Who proposed to an improved sexual system of classification?
(a) B. de Jussieu (b) de Candolle
(c) Michel Adanson (d) Robert Brown

40. a very reasonable natural system of classification proposed by which one of the following great French botanist?
(a) Carolus Linnaeus (b) de Candolle
(c) Michel Adanson (d) Robert Brown

41. A. P. Candolle's work was published in which one of the following?
(a) Families des Plantes (b) Genera Plantarum
(c) Systema Naturae (d) Theorie elementaire

42. Which one of the following person concern with an important discovery of the distinction between Angiosperms and Gymnosperms?
(a) John Lindley (b) Michel Adonson
(c) Robert Brown (d) None of these

43. One who of the following published to Introduction to the Natural order of Plants?
(a) Henri Baillon (b) John Lindley
(c) Robert Brown (d) Stephen Endlicher

44. Thallophyta and Cormophyta are two main groups for Plant Kingdom, were divided by which following concern persons?
(a) John Lindley (b) Robert Brown
(c) Stephen Endlicher (d) None of these

45. The useful descriptions of families and genera of flowering plants are in 'Histoire des Plantes' was published by which one of the following persons?
(a) Bentham (b) Henri Baillon
(c) Hofmeister (d) Hooker

46. In the Origin of Species, famous 'doctrine of evolution' of Charles Darwin published by which researches of the following?
(a) Bentham (b) Hofmeister (c) Hooker (d) None of these

47. Bentham and Hooker two English Botanists carried their researches at great herbarium of which one of the following garden?
(a) Botanical garden (b) National Botanical Garden
(c) Royal Botanical Garden (d) None of these

48. A highly improved system of plant classification given by Bentham and Hooker was published in which one of the following?
(a) Families des Plantes (b) Genera Plantarum
(c) Origin of species (d) Theorie Elementaire

49. Phylogenetic system of classification was first published as guide to the botanical garden of Breslau of Germany by
(a) Bentham and Hooker (b) Eichler
(c) Engler (d) Engler and Prantl

50. For the identification of the whole plant Kingdom, Monumental work called Die Naturlichen Pflanzebfamilien. This was based on which one of the following systems?
(a) Families affinities (b) Flowering system
(c) Natural system (d) Phylogenetic system

51. According to the system in which plant families were arranged according to the increasing complexity of the flower, fruit and seed development was given by which one of the following?
(a) Bentham and Hooker (b) Brown and Lindley
(c) Candolle and Adanson (d) Engler and Prantl

52. The "Classification of Flowering Plants" was written by
(a) Bessey (b) Hallier
(c) Hutchinson (d) Rendle

53. An English botanist, John Hutchinson has given the latest information about the phylogenetic classification of angiosperms published in which one of the following?
(a) Angiosperms
(b) Families of Flowering Plants
(c) Flowering plants
(d) The Classification of Flowering Plants

54. Herbaceae and Lignosae are divided two groups of Angiopsperms suggested by which one of the following persons?
(a) Bessey (b) Hutchinson
(c) Linnaeus (d) Rendle

55. Which one of the following persons suggested that seed plants were divided into two phyla I Gymnospermae and II Angiospermae.
(a) Bessey (b) Hutchinson
(c) Linnaeus (d) Rendle

56. The sub phyla Dicotyledones and Monocolyledones are division of phyla
(a) Angiosperm (b) Bryophytes
(c) Gymnosperm (d) Thallophytes

57 Lignosae and Herbaceae are the division of sub phylum
(a) Dicots (b) Dicotyledons
(c) Monocots (d) Monocotyledones

58. Calyciferae, Corolliferae and Glumiferae are the division of sub phylum
(a) Dicots (b) Dicotyledones
(c) Monocots (d) Monocotyledons

59. In modern trend, the importance to which approach should be given in taxonomy
(a) Biochemical and Molecular
(b) Cytogenetic and Genetic
(c) Embryologic and Phenologic
(d) Morphological and Anatomical

60. Which one of the following is the most recent addition to botanists for the method of grouping of plants?
(a) Biochemical (b) Cytogenetic
(c) Genetic (d) Molecular

61. At the Royal Botanical Gardens, a team led by scientists recently. Kew have devised a new classification of flowering plant families entirely on difference between which one of the following?
(a) Genes (b) Isozymes
(c) Phytochemicals (d) Protein banding

62. The first unit of classification is which one of the following as per International rules of botanical nomenclature?
(a) Class (b) Kingdom
(c) Order (d) Phylum

63. Which one of the following is the next unit of plant classification after kingdom?
(a) Class (b) Family
(c) Order (d) Phylum or Division

64. **After phylum, next unit of plant classification is which one of the following?**

(a) Class (b) Family
(c) Genus (d) Order

65. **The unit of class in the classification is followed by which one of the following?**

(a) Class (b) Family
(c) Order (d) Super order

66. **Which one of the following is the next unit of plant classification after order?**

(a) Family (b) Order
(c) Phylum (d) Class

67. **The next ranking unit of classification after family is which one of the following?**

(a) Genus (b) Phylum
(c) Species (d) Tribe

68. **Genus is the unit of classification which comes in rank after one of the following?**

(a) Division (b) Order
(c) Species (d) Tribe

69. **The last ranking unit of plant classification is which one of the following?**

(a) Class (b) Family
(c) Order (d) Species

70. **The plants which produce seeds are called as one of the following?**

(a) Bryophyta (b) Embryophyta
(c) Lycophyta (d) Spermetophyta

71. **Which one of the following refers to the plants which produces seeds without flower?**

(a) Angiosperms (b) Cryptogams
(c) Embryophytes (d) Gymnosperms

72. **The plant which produce flower and seeds both is known as which one of the following?**

(a) Angiosperm (b) Bryophytes
(c) Gymnosperm (d) Pteridophytes

73. **Angiosperms are Synonym of ________?**

(a) Anthophyta (b) Mangoliophyta
(c) Spermatophytes (d) Both (b) and (c)

74. In the plant classification, for dicotyledones class which one of the following botanical term is used?
(a) Anthophyta (b) Liliopsida
(c) Magnoliophyta (d) Magnoliopsida

75. In the plant classification, for monocotyledones class which one of the following botanical term is used?
(a) Anthophyta (b) Liliopsida
(c) Magnoliophyta (d) Magnoliopsida

76. The binomial nomenclature of any plant includes which one of the following?
(a) Class and Subclass (b) Family and Order
(c) Genus and Species (d) Tribe and Sub-tribe

77. A modern system of pant taxonomy the APG II system was published by Angiosperm phylogeny group (APG)
(a) 2001 (b) 2003
(c) 2004 (d) 2005

78. Now these days family Palmae is known as one of the following?
(a) Arecaceae (b) Clusiaceae
(c) Fabaceae (d) Poaceae

79. At present, family Gramineae is known as one of the following?
(a) Arecaceae (b) Clusiaceae
(c) Fabaceae (d) Poaceae

80. Family Cruceferae, now these days known as___________?
(a) Arecaceae (b) Brassicaceae
(c) Clusiaceae (d) Poaceae

81. Now these days, family Leguminosae is known by another name as which one among the following?
(a) Arecaceae (b) Fabaceae
(c) Lamiaceae (d) Poaceae

82. Family Guttiferae name is changed as which one of the following?
(a) Clusiaceae (b) Fabaceae
(c) Lamiaceae (d) Poaceae

83. Umbelliferae family is known now these days as one of the following?
(a) Apiaceae (b) Asteraceae
(c) Fabaceae (d) Lamiaceae

84. Family Labiatae is now known as which one of the following?
(a) Apiaceae (b) Arecaceae
(c) Asteraceae (d) Lamiaceae

85. Compositae family is now changed as which one of the following?
(a) Apiaceae (b) Arecaceae
(c) Asteraceae (d) Poaceae

86. Which of the following species is restricted to a specific area?
(a) Allopatric species (b) Endemic species
(c) Sibling species (d) Sympatric species

ANSWERSHEET

5. Plant Classification

1. d	2. c	3. a	4. d	5. b	6. d	7. d
8. d	9. c	10. d	11. d	12. d	13. c	14. b
15. a	16. b	17. a	18. c	19. d	20. c	21. d
22. a	23. a	24. a	25. a	26. d	27. c	28. b
29. a	30. a	31. d	32. b	33. a	34. d	35. d
36. c	37. b	38. d	39. a	40. b	41. b	42. c
43. b	44. c	45. c	46. b	47. c	48. b	49. c
50. d	51. d	52. d	53. c	54. b	55. b	56. a
57. b	58. d	59. b	60. c	61. a	62. b	63. d
64. a	65. c	66. a	67. d	68. d	69. d	70. d
71. d	72. a	73. d	74. d	75. b	76. c	77. b
78. a	79. d	80. b	81. b	82. a	83. a	84. d
85. c	86. b					

6

Botanical Description

1. A plant that completes several generations in one year is known as which one of the following?
(a) Annual (b) Biennial
(c) Ephemeral (d) Perennial

2. A plant that completes its life cycle with in single season is known as which one of the following?
(a) Annual (b) Biennial
(c) Ephemeral (d) Perennial

3. A plant that completes its life cycle in two years is known as which one of the following?
(a) Annual (b) Biennial
(c) Ephemeral (d) Perennial

4. A plant that continues its growth and reproduces for many years is known as which one of the following?
(a) Annual (b) Biennial
(c) Ephemeral (d) Perennial

5. Organic substances transported in the phloem can be demonstrated by as which one of the following?
(a) Defoliation (b) Grafting
(c) Ringing the stem (d) Root pressure

6. When thalamus grows upward and makes copular structure but free from ovary on which remaining whorls are situated in periphery such condition is known as which one of the following?
(a) Epigynous (b) Hypogynous
(c) Perigymous (d) Zygomorphic

7. A non woody, small usually annual plant is called as which one of the following?
(a) Climber (b) Creaper
(c) Herb (d) Shrub

8. A woody plant of relatively low height, branching near the base and do not have main trunk is known as which one of the following?
   (a) Climber (b) Herb
   (c) Shrub (d) Tree
9. A perennial woody plant with one trunk is known as which one of the following?
   (a) Creaper (b) Shrub
   (c) Trailer (d) Tree
10. A plant with weak stem and with special organs which help it ascend over neighbouring objects is called as which one of the following?
   (a) Climber (b) Creaper
   (c) Trailer (d) Twinner
11. A plant with weak and thin stem which ascend by coiling round its support and do not develop any special organ for this purpose is called as which one of the following?
   (a) Climber (b) Creaper
   (c) Trailer (d) Twinner
12. A plant with weak horizontal shoot that roots throughout most of its length is called as which one of the following?
   (a) Climber (b) Creaper
   (c) Trailer (d) Twinner
13. A plant with weak and thin stem lying prostrate on the ground but lack rooting is called as which one of the following?
   (a) Climber (b) Creaper
   (c) Trailer (d) Twinner
14. VAM is which one of the following?
   (a) Bioherbicide (b) Bioinsecticide
   (c) Ectomycorrhiza (d) Endomycorrhiza
15. A plant growing in or near water is called as which one of the following?
   (a) Epiphytic (b) Hydrophytic
   (c) Hydroponic (d) Xerophytic
16. Microsporogenesis refers to the formation of microspores inside the microsporangia or pollen sacs of the seed plants, occurs in which one of the following part of plants?
   (a) Anthers (b) Ovules
   (c) Roots (d) Stems

17. A plant growing in dry environment is known as which one of the following?
(a) Drought resistant (b) Epiphytic
(c) Hydrophytic (d) Xerophytic

18. A plant growing on another plant for support but does not take nurishment is known as which one of the following?
(a) Climber (b) Epiphytic
(c) Parasite (d) Xerophytic

19. Polyadelphous situation found in of which one of the following family?
(a) Labiatae (b) Ligumenaceae
(c) Liliaceae (d) Rutaceae

20. A plant that grows on another organism and obtains nourishment is known as which one of the following?
(a) Epiphytic (b) Parasitic
(c) Thallophytic (d) Xerophytic

21. A primary root develops from radicle of the embryo and growing indefinitely is known as which one of the following?
(a) Fusiform (b) Napiform
(c) Parasitic (d) Taproot

22. A roots arising from other parts like stem or leaf other than the radicle are called as which one of the following?
(a) Adventitious root (b) Fusiform
(c) Napiform (d) Tap root

23. A stem having filled interior is called as which one of the following?
(a) Angular (b) Clum
(c) Fistular (d) Solid

24. A stem having a hollow interior is called as which one of the following?
(a) Angular (b) Clum
(c) Fistular (d) Solid

25. Example of day neutral plant is which one of the following crops?
(a) Barley (b) Rice
(c) Tomato (d) Wheat

26. A stem covered with hairs is known as which one of the following?
(a) Glabrous (b) Hairy
(c) Pubescent (d) Spiny

27. A stem without hair is known as which one of the following?
(a) Glabrous (b) Hairy
(c) Pubescent (d) Spiny

28. A stem covered with prickly spines is called as which one of the following?
(a) Glabrous (b) Hairy
(c) Pubescent (d) Spiny

29. A leaf with one blade which may be entire or incised to any depth but not up to the mid rib or petiole is called as which one of the following?
(a) Compound (b) Simple
(c) Ternate (d) Whorled

30. Plants that flower only once in their life span is called as which one of the following?
(a) Cleistocarpic (b) Monocarpic
(c) Pericarpic (d) Polycarpic

31. A leaf in which the blade is incised up to the mid rib or petiole into two or more leaflets is which one of the following?
(a) Compound (b) Multipinnate
(c) Paripinnate (d) Pinnate

32. Long, slender, needle shaped, cylindrical, shape of leaf is known as which one of the following?
(a) Acicular (b) Lanceolate
(c) Linear (d) Oblong

33. Long, narrow and flat, sides nearly parallel cylindrical, shape of leaf is known as which one of the following?
(a) Acicular (b) Lanceolate
(c)Linear (d) Oblong

34. Which of the following is a living simple tissue involved as strengthening tissue for the plant body?
(a) Collenchymas (b) Parenchyma
(c) Sclerenchyma (d) Both (a) and (c)

35. Biconvex lens shaped leaf is called as which one of the following?
(a) Acicular (b) Lanceolate
(c) Linear (d) Oblong

36. Long and broad with the margin nearly parallel shape of leaf is known as which one of the following?
(a) Acicular (b) Lanceolate
(c) Linear (d) Oblong

37. **Oval but slightly broader at the base and tapering towards the apex shape of leaf is known as which one of the following?**
(a) Lanceolate (b) Linear
(c) Oblong (d) Ovate

38. **Oval but slightly broader at the apex and tapering towards the base shape of leaf is known as which one of the following?**
(a) Oblong (b) Obovate
(c) Ovate (d) Round

39. **Which of the following subcellular components is likely to be absent in a plant root?**
(a) Chloroplasts (b) Large vacuoles
(c) Leucoplasts (d) Mitochondria

40. **Leaf blade situated on stalk or petiole is called as which one of the following?**
(a) Petiolate (b) Phyllode
(c) Sessile (d) Sub sessile

41. **Leaf having very short petiole is called as which one of the following?**
(a) Grooved (b) Sessile
(c) Sub petiolate (d) Sub sessile

42. **Leaf without a petiole or stalk is which one of the following?**
(a) Petiolate (b) Phyllode
(c) Sessile (d) Sub petiolate

43. **Flowering in plant is controlled by which one of the following?**
(a) Floriculture (b) Photoperiodism
(c) Transpiration Rate (d) Vernalization

44. **Leaf having petiolate flat possessing lateral extensions is called as which one of the following?**
(a) Phyllode (b) Sessile
(c) Spong (d) Winged

45. **Petiole modified into leaf like structure is known as which one of the following?**
(a) Adnate (b) Ligulate
(c) Phyllode (d) Scaly

46. **A leaf surface smooth but dull is called as which one of the following?**
(a) Glabrous (b) Glaucous
(c) Mealy (d) Pubescent

47. A leaf surface covered with whitish substance that rubs off exposing green and shining surface is called as which one of the following?
(a) Glabrous (b) Glaucous
(c) Sticky (d) Waxy

48. A leaf surface covered by spines called as which one of the following?
(a) Rough (b) Spiny
(c) Thorny (d) Trichomous

49. A leaf surface covered with short and smooth hairs called as which one of the following?
(a) Hairy (b) Pubescent
(c) Rough (d) Spiny

50. A leaf surface covered with various types of hair is known as which one of the following?
(a) Hairy (b) Pubescent
(c) Rough (d) Spiny

51. In a bisexual flower when gynoecium matures earlier than androecium it is called which one of the following?
(a) Autogamy (b) Heterogamy
(c) Protandry (d) Protogamy

52. The leaf having veinlets irregularly distributed forming a net work such venation are called as which one of the following?
(a) Convergent (b) Divergent
(c) Multicostate (d) Reticulate

53. A leaf have single prominent mid rib such venation is known as which one of the following?
(a) Convergent (b) Parellel
(c) Reticulate (d) Unicostate

54. A leaf have more than one prominent veins arising at the apex of the petiole such venation is known as which one of the following?
(a) Convergent (b) Divergent
(c) Multicostate (d) Reticulate

55. Veins run parallel to each other and net work is not formed such venation is known as which one of the following?
(a) Convergent (b) Divergent
(c) Multicostate (d) Parallel

56. The channels of protoplasm that exists through cell walls that connect one cell to another are called.
(a) Cytoplasmic threads (b) Plasmodesmata
(c) Stomata (d) Watsamata

57. Leaf texture leathery in touch is known as which one of the following?
(a) Cereous (b) Coriaceous
(c) Scarious (d) Succulent

58. Leaf texture thin, flexible and translucent like a membrane called as which one of the following?
(a) Cereous (b) Membranous
(c) Scarious (d) Translucent

59. Leaf texture thin, dry and membranous often non green called as which one of the following?
(a) Cereous (b) Coriaceous
(c) Membranous (d) Scarious

60. Leaf texture thick, fleshy and juicy known as which one of the following?
(a) Cereous (b) Coricaceous
(c) Scerious (d) Succulent

61. Spores produced as a result of budding is termed as which one of the following?
(a) Arthrospores (b) Blastospores
(c) Conidiospores (d) Chlamydospores

62. A leaf texture is waxy called as which one of the following?
(a) Cereous (b) Coriaceous
(c) Scarious (d) Waxy

63. In florescence with monopodial branching is known as which one of the following?
(a) Corymb (b) Cymose
(c) Receme (d) Recemose

64. Major food crops of the world belong to family of one among the following?
(a) Leguminosae (b) Solanaceae
(c) Cruciferae (d) Gramineae

65. Recemose inflorescence having common axis and stalked flowers arranged in acropetal succession is called as which one of the following?
(a) Corymb (b) Cymose
(c) Panicle (d) Receme

66. If axis of receme is branched it is called as which one of the following?
(a) Catkin (b) Panicle
(c) Spike (d) Spikelet

67. Inflorescence having common axis but flowers are sessile such recemose is known as which one of the following?
(a) Catkin (b) Panicle
(c) Spike (d) Spikelet

68. Very small spike of a compound spike forming a unit is called as which one of the following?
(a) Calkin (b) Panicle
(c) Spike (d) Spikelet

69. The modified receme with relatively a short main axis in which lower flowers have much elongated pedicels so that flowers come to the same level called as which one of the following?
(a) Capitulum (b) Compound umbel
(c) Corymb (d) Umbel

70. The modified receme in which stalk of flowers have same length and they seem to arise from same point is called as which one of the following?
(a) Capitulum (b) Compound umbel
(c) Corymb (d) Umbel

71. A dense concave, convex, spherical or flat topped inflorescence of numerous sessile flowers clustered on common disc such recemose is known as which one of the following?
(a) Capitulum (b) Compound umbel
(c) Corymb (d) Umbel

72. An inflorescence with sympodial branching the main axis terminates into flower and lateral branches arise below it which also terminates into flowers and central flower open first is called as which one of the following?
(a) Corymb (b) Cymose
(c) Recemose (d) Umbel

73. The main axis of cymose ending in a flower producing only one lateral branch at a time each terminate into a flower is known as which one of the following?
(a) Biparous (b) Glomerule
(c) Multiparous (d) Uniparous

74. The main axis of cymose ending in a flower producing a pair of lateral branches which too terminate into flower is called as which one of the following?
(a) Biparous (b) Glomerule
(c) Multiparous (d) Uniparous

75. The main axis of cymose ending in a flower producing more than two lateral branches which too terminate into flowers called as which one of the following?
(a) Biparous (b) Glomerule
(c) Multiparous (d) Uniparous

76. A very dense single cyme with central flower blooming first is called as which one of the following?
(a) Biparous (b) Glomerule
(c) Multiparous (d) Uniparous

77. A flower having all the four whorls is called as which one of the following?
(a) Complete (b) Hermaphrodite
(c) Palygamous (d) Perfect

78. A flower lacking one or more whorls is known as which one of the following?
(a) Imperfect (b) Incomplete
(c) Monocecious (d) Unisexual

79. A flower having both male and female organs is known as which one of the following?
(a) Dioecious (b) Hermaphrodite
(c) Perfect (d) Both (b) and (c)

80. A flower having only one reproductive whorl is known as which one of the following?
(a) Imperfect (b) Monoecious
(c) Unisexual (d) Both (a) and (c)

81. A plant having more than two types of flower is called as which one of the following?
(a) Dioecious (b) Hermaphrodite
(c) Monoecious (d) Polygamous

82. The science which deals with the study of the external structure or physical form of plants is known as which one of the following?
(a) Plant morphology (b) Plant physiology
(c) Plant taxonomy (d) None of these

83. A plant having both staminate and pistillate flowers is known as which one of the following?
(a) Dioecious (b) Hermaphrodite
(c) Monoecious (d) Polygamous

84. A plant having either staminate or pistillate flowers is known as which one of the following?
(a) Dioecious (b) Hermaphrodite
(c) Monoecious (d) Polygamous

85. The flower in which ovary is situated on top of thalamus and remaining three whorls below the ovary such flower is called as which one of the following?
(a) Epigynous (b) Hypogynous
(c) Perigymous (d) Zygomorphic

86. When three whorls are situated above the ovary as the thalamus grows upword surrounding ovary this condition is called as which one of the following?
(a) Epigynous (b) Hypogynous
(c) Perigymous (d) Zygomorphic

87. Dry and scaly outer bracts at the base of an individual spikelet are known as which one of the following?
(a) Bracteoles (b) Glume
(c) Lemma (d) Palea

88. The outer bract of the floret may bear awn is called as which one of the following?
(a) Bracteoles (b) Glume
(c) Lemma (d) Palea

89. The inner bract of the floret is called as which one of the following?
(a) Bracteoles (b) Glume
(c) Lemma (d) Palea

90. All the stamens of a flower are united in two bundles by fusion of their filaments is called as which one of the following?
(a) Diadelphous (b) Monoadelphous
(c) Polyadelphous (d) Syngenesious

91. All stamens of a flower are united in two bundles this situation is called as which one of the following?
(a) Diadelphous (b) Monoadelphous
(c) Polyadelphous (d) Syngenesious

92. All the stamens unite to form more than two groups this situation is known as which one of the following?
(a) Diadelphous (b) Monoadelphous
(c) Polyadelphous (d) Syngenesious

93. Stamens are indefinite and free this situation is called as which one of the following?
(a) Diadelphous (b) Monoadephous
(c) Polyadelphous (d) Polyandrous

94. A bisexual flower which never opens in its life span is known as which one of the following?
(a) Cleistogamous (b) Heterogamous
(c) Homogamous (d) Polygamous

95. Ovule integument gets transformed into which one of the following?
(a) Cotyledons (b) Fruit wall
(c) Seed (d) Seed coat

96. The Gynoecium comprised of a single carpel is known as which one of the following?
(a) Monocarpel (b) Monocarpellary
(c) Single carpellary (d) Unicarpellary

97. The gynoecium comprised of two carpels is known as which one of the following?
(a) Bicarpellary (b) Bicarpelled
(c) Dicarpellary (d) Two carpelled

98. The gynoecium comprised of five carpels is known as which one of the following?
(a) Five carpelled (b) Multicarpellary
(c) Pentacarpellary (d) Polycarpellary

99. The gynoecium comprised of many carpels is knwon as which one of the following?
(a) Five carpelled (b) Multicarpellary
(c) Pentacarpellary (d) Polycarpellary

100. Protandy coupled with herkogamy enhances the opportunity for cross-pollination. Fertilized flowers develop into capsules, each with hundreds of small seeds. Protandy is a common phenomenon of which one of the following family?
(a) Asteraceae (b) Crusifereae
(c) Cucurbitaceae (d) Malvaceae

ANSWERSHEET

6. Botanical Description

1. c	2. a	3. b	4. d	5. a	6. c	7. c
8. c	9. d	10. a	11. d	12. b	13. b	14. d
15. b	16. a	17. d	18. b	19. d	20. b	21. d
22. a	23. d	24. c	25. c	26. b	27. a	28. d
29. b	30. b	31. a	32. a	33. c	34. a	35. b
36. d	37. a	38. b	39. a	40. a	41. b	42. c
43. b	44. d	45. c	46. a	47. b	48. b	49. b
50. a	51. d	52. d	53. d	54. c	55. d	56. b
57. b	58. b	59. d	60. d	61. b	62. a	63. d
64. d	65. d	66. b	67. c	68. d	69. c	70. d
71. a	72. b	73. d	74. a	75. c	76. b	77. a
78. b	79. d	80. d	81. d	82. a	83. c	84. a
85. b	86. a	87. b	88. c	89. d	90. b	91. a
92. c	93. d	94. a	95. d	96. b	97. a	98. c
99. b	100. d					

7

Economic Field Crops

1. The world's major food crops belong to which one of the following families?
   (a) Cruciferae (b) Gramineae
   (c) Leguminosae (d) Solanaceae
2. The prominent food crops belong to which one of the following?
   (a) Cereals (b) Oilseeds
   (c) Pulses (d) Vegetables
3. In India, which one of the following cereals is grown largest?
   (a) Pearl millet (b) Rice
   (c) Sorghum (d) Wheat
4. Which one of the following plant is a natural insecticide?
   (a) Cinerin (b) Nicotine
   (c) Pyrethrum (d) All of the above
5. Saffron (*Crocus sativus*) commonly known as the saffron crocus, belongs to which one of the following family?
   (a) Ephedraceae (b) Iridaceae
   (c) Liliaceae (d) Poaceae
6. Saffron is a spice, produced from which one of the following?
   (a) Stamens of *Crocus* (b) Stigma of *Crocus*
   (c) Style of *Crocus* (d) Both (b) and (c)
7. Which one of the following is the most important cereal grain crop in the world?
   (a) Corn (b) Rice
   (c) Sorghum (d) Wheat
8. Coir of commerce comes from which part of coconut?
   (a) Epicarp (b) Mesocarp
   (c) Seedcoat (d) Endocarp

9. The saffron crocus is a __________ that is self-incompatible and male sterile?
(a) Monopoloid (b) Triploid
(c) Dipoloide (d) Hexapoloide

10. In saffron, glucosidepicrocrocin is responsible for which one of the following?
(a) Flavour (b) Taste
(c) Both (a) and (b) (d) None of these

11. The major source of sugar is
(a) Date palm (b) Stevia
(c) Sugar beet (d) Sugarcane

12. The most important vegetable of world diet is
(a) Brinjal (b) Potato
(c) Tomato (d) Vegetable pea

13. Which one of the following plants has medicinal importance?
(a) *Brassica oleraceae* (b) *Coffea robusta*
(c) *Cryptostegia grandiflora* (d) *Rauwolfia serpentina*

14. Medicinal plant sarpagandha is a flowering plant, belongs to which one of the following families?
(a) Apocynaceae (b) Gelsemiaceae
(c) Loganiaceae (d) Rubiaceae

15. For cultivated species of rice, which one of the following is true regarding tillering?
(a) INDICA = Profuse (b) JAPONICA = Medium
(c) JAVANICA = Low (d) All of the above

16. The native place of Hevea rubber is which one of the following?
(a) Asia minor (b) India
(c) Malasia (d) Mexico

17. In the world, which one of the following is largest industrial crop?
(a) Cotton (b) Jute
(c) Soybean (d) Sugarcane

18. Which one of the following type of cotton has 90% of world production?
(a) *Gossypium arboreum* (b) *Gossypium barbadense*
(c) *Gossypium herbaceum* (d) *Gossypium hirsutum*

19. Which one of the following type known as extra-long staple cotton?
(a) *Gossypium arboreum* (b) *Gossypium barbadense*
(c) *Gossypium herbaceum* (d) *Gossypium hirsutum*

20. Which one of the following corn is characterized by a translucent and wrinkled appearance upon drying, and a sweet taste when immature?
(a) *Z. mays indentata* (b) *Z. mays saccharata*
(c) *Z. mays tunicate* (d) None of these

21. The major use of sugarcane including sugar is?
(a) Alcohol (b) Biofuel
(c) Both (a) and (b) (d) None of these

22. Reserpine is analkaloidthat has been used as drug for the control of high blood pressure. This chemical compound is extracted from which of the following plants?
(a) *Atropa belladonna* (b) *Brassica oleraceae*
(c) *Digitalis purpurea* (d) *Rauwolfia serpentine*

23. Textile industries mainly based on which one of the following economic crops?
(a) Cotton (b) Jute
(c) Both (a) and (b) (d) None of these

24. Which one of the following is true regarding drought for cultivated species of rice?
(a) INDICA = Resistant (b) JAPONICA = Intermediate
(c) JAVANICA = Sensitive (d) All of the above

25. A milk preparation can be made from the seeds of on which one of the following crops?
(a) Barley (b) Gram
(c) Grapes (d) Soyabean

26. Fibers derived from epidermis of which one of the following fibrous crops that have commercial importance?
(a) Cotton (b) Flax (c)Hemp (d) None of these

27. A huge foreign exchange is earned by which one of the following oilseed crops?
(a) Groundnut (b) Mustard
(c) Sesame (d) Soybean

28. Which one of the following seeds are good sources can be made milk like preparation?
(a) Barley (b) Gram (c) Grapes (d) Soybean

29. One of the plants introduced from old world to new world is which one of the following?
(a) Maize (b) Potato
(c) Rice (d) Sweet potato

30. Soybean seeds content account for ______ proteins.
(a) 20% (b) 40%
(c) 60% (d) None of these

31. What percentage of oils content in soybean seeds?
(a) 20% (b) 40%
(c) 60% (d) None of these

32. Source of rubber is which one of the following?
(a) *Cedrus depdara* (b) *Hevea brasilensis*
(c) *Michelia champaca* (d) *Tectona grandis*

33. Saffron is produced from which one of the following?
(a) Style and stigma of candytuft (b) Style and stigma of crocus
(c) Style and stigma of Hibiscus (d) Style and stigma of poppy

34. The pulse crop which has diverse use in daily diet of India?
(a) Chickpea (b) Green gram
(c) Pea (d) Pigeon pea

35. Which one of the following corns is available of the multiple colors, the yellow or white kernels dominate commercial production?
(a) *Z. mays indentata* (b) *Z. mays saccharata*
(c) *Z. mays tunicate* (d) None of these

36. By which one of the following the flax fibers are obtained?
(a) *Cannabis sativa* (b) *Cocos nucifera*
(c) *Crotolaria juncea* (d) *Linum ustistissimum*

37. Acalyphine is extracted from annual erect herb is one of the following?
(a) *Acalypha indica* (b) *Aegle marmelos*
(c) *Cissus quadrangularis* (d) *Mimosa pudica*

38. *Acalyphaindica*belong to which one of the following family?
(a) Achariaceae (b) Euphorbiaceae
(c) Euphroniaceae (d) Turneraceae

39. Saponins are extracted from which one of the following plants?
(a) *Solanum melongena* (b) *Solanum nigrum*
(c) *Solanum tuberosam* (d) None of these

40. Bio-diesel is obtained from which one of the following crops?
(a) Groundnut (b) Rice (c) Soybean (d) Wheat

41. Which one of the following are true statements?
(a) Alcoholic beverages are distilled from the grains of rice.
(b) Cotton seeds cake is used as good organic manure.
(c) Sake is an alcoholic beverage in Japan, prepared by the fermentation of rice.
(d) All of the above

42. Which one of the following is correct about cotton (Gossypium sps)?
(a) *G. arboreurn* = Indian cotton
(b) *G. barbadense* = Egyption / sea island cotton
(c) *G. hirsutum = American / upland cotton*
(d) All of the above

43. Which one of the following is the centre of origin for cotton?
(a) Central Africa (b) North Africa
(c) South Africa (d) West Africa

44. Which one of the following is correct with respect to the distribution of rice, except?
(a) GLABERRIMA = Widely distributed to America
(b) INDICA = Tropical e.g. Asia India
(c) JAPONICA = Sub-tropical to warm temperate e.g. Japan
(d) JAVANICA = Intermediate Sub-tropical e.g. Indonesia

45. In India, sunflower (*Helianthus annuus*) was introduced from USSR in which one of the following years?
(a) 1960 (b) 1964 (c) 1965 (d) 1969

46. Match the following crops with their center of origins.

1. Maize	A. Africa (According to vavilor), India (According to Decandole)
2. Wheat	B. Central America
3. Rice	C. India
4. Finger millet	D. S. W. Asia
5. Mustard	E. S. E. Asia

(a) 1B 2D 3E 4A 5C (c) 1C 2A 3E 4D 5B
(b) 1E 2A 3D 4C 5B (d) 1C 2D 3A 4E 5B

47. Match the following cereals crop with their center of origins.

1. Diploid (2n=14) Wheat	A. Asia minor
2. Tetraploid (2n=28) Wheat	B. S. E. Africa
3. Hexaploid (2n=42) Wheat	C. W. Africa
4. Sorghum	D. Central Asia
5. Pearl millet	E. Abyssinia, North Africa

(a) 1A 2E 3D 4B 5C (c) 1E 2B 3A 4C 5D
(b) 1A 2E 3B 4C 5D (d) 1B 2E 3A 4D 5C

48. Match the following oil seed crops with their center of origins.

1. Ground nut	A. Brazil
2. Sesame	B. India, and Ethopia (Afric(a)
3. Sunflower	C. America
4. Safflower	D. Africa and Afghanistan
5. Castor	E. Africa

(a) 1B 2D 3E 4A 5C (c) 1C 2A 3E 4D 5B
(b) 1E 2A 3D 4C 5B (d) 1C 2D 3A 4E 5B

49. Match the following pulses crops with their center of origins.

1. Red gram	A. China
2. Soybean	B. India
3. Bengal gram	C. India
4. Green gram	D. S. W. Africa and Mediterranean region (According to Vavilov, 1926), Turkey and Syria (According to Vander Maesun, 1984)
5. Black gram	E. Africa and India

(a) 1B 2D 3E 4A 5C (c) 1C 2A 3E 4D 5B
(b) 1E 2A 3D 4C 5B (d) 1C 2D 3A 4E 5B

50. Match the following *Brassica* spp. with their characteristics.

1. *Brassica compestris*	A. Black mustard
2. *Brassica napus*	B. Indian mustard
3. *Brassica nigra*	C. Indian rape seed
4. *Brassica alba*	D. European rape seed
5. *Brassica juncea*	E. White mustard

(a) 1B 2D 3E 4A 5C (c) 1C 2A 3E 4D 5B
(b) 1E 2A 3D 4C 5B (d) 1C 2D 3A 4E 5B

51. Match the following center of cultivations of genus *Oryza* with their botanical name.

1. Asia	A. *O. sativa, O. nivara, O. rufipogan, O. minuta, O. punctata, O. officinalis, O. granulata, O. meyeriane*
2. Africa	B. *O. longistaminata, O. glaberrima, O. barthii, O. eichingeri, O. brachantha*
3. America	C. *O. glumaepatula, O. grandiglumis, O. latifolia, O. alata*
4. Australia	D. *O. meridionalis, O. australiensis*
5. New Guninea	E. *O. longiglumis, O.schlechter*

(a) 1A 2E 3D 4B 5C (c) 1E 2B 3A 4C 5D
(b) 1A 2E 3B 4C 5D (d) 1B 2E 3A 4D 5C

52. Match the following common name with their genus *Triticum*.

1. Wild einkorn — A. *T. dicoccoides*
2. Einkorn — B. *T. boeticum*
3. Wild Emmer — C. *T. durum*
4. Emmer — D. *T. dicoccum*
5. Macaroni wheat — E. *T. aegilopoides, T .monococum*

(a) 1A 2E 3D 4B 5C
(b) 1A 2E 3B 4C 5D
(c) 1E 2B 3A 4C 5D
(d) 1B 2E 3A 4D 5C

53. Match the following genus of *Triticum* with their common name.

1. *T. aestivum* — A. Dwarf wheat
2. *T. compactum* — B. Club wheat
3. *T. sphaerococcum* — C. Spelt wheat
4. *T. spelta* — D. Macha wheat
5. *T. macha* — E. Common or bread wheat

(a) 1A 2E 3D 4B 5C
(b) 1A 2E 3B 4C 5D
(c) 1E 2B 3A 4C 5D
(d) 1B 2E 3A 4D 5C

ANSWERSHEET

7. Economic Field Crops

1. b	2. a	3. b	4. d	5. b	6. d	7. d
8. b	9. b	10. a	11. d	12. b	13. d	14. a
15. d	16. c	17. a	18. d	19. a	20. b	21. c
22. d	23. a	24. d	25. d	26. a	27. d	28. d
29. c	30. b	31. a	32. b	33. b	34. a	35. a
36. d	37. a	38. b	39. b	40. b	41. d	42. d
43. a	44. a	45. d	46. a	47. a	48. c	49. b
50. d	51. b	52. d	53. c			

8

Plant Hormones

1. In the context to hormone, which one of the following is a low molecular weight organic compound?
(a) Carbohydrates (b) Enzymes
(c) Vitamins (d) None of these
2. Hormones are those organic substances that
(a) Act as enzyme or coenzyme
(b) Influence synthesis of enzymes
(c) Both (a) and (b) (d) None of these
3. The site of hormones synthesis is known as which one of the following?
(a) Cytosols (b) Effectors
(c) Torgets (d) None of these
4. Which one of the following is known as the site of harmones action?
(a) Cytosols (b) Effectors
(c) Torgets (d) None of these
5. The word hormone is derived from which one of the following languages, meaning *set in motion*?
(a) Germen (b) Greek
(c) Latin (d) None of these
6. Plant hormones affect which one of the following?
(a) Gene expression (b) Transcription levels
(c) Cellular division and growth (d) All of the above
7. Plant hormones are naturally produced within plants, though very similar chemicals are produced by which one of the following organism that can also affect plant growth?
(a) Fungi (b) Bacteria
(c) Both (a) and (b) (d) None of these

8. Very often at sites of active growth within which one of the following, the production of plant hormones occurs, before cells have fully differentiated?
(a) Cambium (b) Meristems
(c) Xylem (d) None of these

9. At which one of the following very low concentrations of hormones require for plant responses?
(a) 10^{-5} to 10^{-4} mol/L (b) 10^{-6} to 10^{-5} mol/L
(c) 10^{-7} to 10^{-6} mol/L (d) None of these

10. Which one of the following is/are the additional benefits of applying PGRs?
(a) Disease suppression (b) Greener leaves
(c) Less water use (d) All of the above

11. The name hormone was first used by one of the following scientist in 1904.
(a) Eiichi Kurosawa (b) W. Bayliss and E. Starling
(c) Teijiro Yabuta and Sumuki (d) None of these

12. In 1948, which one of the following term was coined by Thimann?
(a) Ethylene (b) Harmone
(c) Phytohormone (d) Animal harmone

13. Which one of the following is true regarding plant hormone which regulates growth, differentiation, or other specific physiological processes?
(a) Synthesized by a plant
(b) It is a detoxification product
(c) It is a nutrient or coenzyme (d) Both (a) and (c)

14. Plant hormones act by affecting the activities of which one of the following?
(a) Enzymes (b) Genes
(c) Genes and Enzymes (d) None of these

15. The auxins are known as the main auxin of higher plants and formed from L-tryptophan?
(a) Indole-3-acetonitrile (b) Indole-3-acetic acid
(c) 1-Naphthaleneacetic acid (d) None of these

16. In shoots, branching is inhibited by which one of the following harmones from the tip of a growing shoot?
(a) Abscisic acid (b) Auxins
(c) Cytokinins (d) Gibberellins

17. Induction of gene expression is played by which one of the following plant hormones?
(a) Abscisic acid
(b) Auxins
(c) Cytokinins
(d) Gibberellin

18. Most potential dormancy inducer is which one of the following?
(a) ABA
(b) Coumarin
(c) Ethylene
(d) Scopoletins

19. In combination with which one of the following, auxin is required for the sustained proliferation of many cultured plant tissues?
(a) Abscics acids
(b) Cytokinins
(c) Ethylene
(d) Gibrelines

20. Which one of the following is a volatile plant hormone?
(a) Auxine
(b) Cytokinins
(c) Ethylene
(d) Gibrelines

21. Brassinolide, plant growth regulators was the first identified brassinosteroid and was isolated from extracts of rapeseed (*Brassica napus*) pollen in which one of the following years?
(a) 1977 (b) 1978 (c) 1979 (d) None of these

22. At the concentration range of 0.1-10.0 mg/L, which one of the following phytohormones is generally used in plant cell culture?
(a) Auxins (b) Cytokinins (c) Polyamines (d) None of these

23. The plant hormones which favour cell division and cell expansion respectively are which one of the following?
(a) Cytokinins and Auxins (b) Cytokinins and ethylene
(c) Gibberellins and Auxins (d) Gibberellins and cytokinins

24. Which is an antiauxine from the following that inhibits the growth regulating function of an auxin?
(a) p-chlorophenoxyisobutyric acid (PCI(B)
(b) 2, 3, 5-Triodobenzoic acid (TIB(A)
(c) 2,4-dichlorophenoxy acetic acid (2,4-(D)
(d) Both (a) and (B)

25. Inhibition of sencsences is a physiological role of ________ plant harmones?
(a) Cytokinines
(b) Ethylinens
(c) Gibberiline
(d) None of these

26. Which one of the following phytohormones is in stimulation of abscission of leaves?
(a) Abscisic acid (AB(A)
(b) Gibrelines (G(A)
(c) Cytokininss (CKs)
(d) Ethylene

27. Dormancy of can be broken by treating the seed with 1% thio-urea + 1 ppm GA_3 for one hour followed by 3% ethylene chlorophydrin solution and storage for 72 hour?
(a) Faba bean
(b) Lentil
(c) Jatropha
(d) Potato

28. Which one of the following phytohormones formed by the degradation of carotenoids?
(a) Abscisic acid (AB(A)
(b) Gibrelines (G(A)
(c) Cytokininss (CKs)
(d) Ethylene

29. Brassinosteroids (BRs) are a class of polyhydroxysteroids that biosynthesized is related with which one of he following in plants?
(a) Campesterol
(b) Cholesterol
(c) Sitosterol
(d) None of these

30. Which one of the following are not correct statements?
(a) Carrot root phloem bioassay is not bioassay method for kinetin.
(b) phosphon-D and Amo-1618 are anti-gibberellins including Cycocel (CC(C).
(c) Cytokinins are known to help retention of chlorophyll.
(d) Richmond-Lang effect pertains to the delay of senescence.

31. Which one of the following are not correct statements?
(a) The concentration of diffusible auxins is maximum in apical buds.
(b) Abscisic acid (AB(A) phytohormones are regarded as antitranspirant.
(c) Both (a) and (b)
(d) None of these

32. Foolish seedling disease in rice with extremely fast growth of spindly and pale seedling, that break off easily is caused by the substance secreted by a parasitic fungi (*Fusarium moniliforme*) is
(a) Abscisic acid
(b) Auxin
(c) Cytokinnins
(d) Gibberellin

33. At the concentration range of 0.01-10.0 mg/L, which one of the following phytohormones is generally used in plant cell culture?
(a) Auxins (b) Cytokinins (c) Polyamines (d) None of these

34. **Which one of the following hormones can actively stimulate RNA synthesis in dwarf mutant plants, making them grow like normal plants?**
(a) Gibrelines (b) Cytokininss
(c) Ethylene (d) None of these

35. **The hormone does not synthesize in the developing seed is which one of the following?**
(a) ABA (b) Auxin
(c) Cytokinin (d) Gibberlin

36. **A plant growth regulator play role to delay of sencsences is?**
(a) Auxines (b) Cytokinines
(c) Ethylinens (d) Gibberiline

37. **Which one of the following hormone is also called as stress hormone?**
(a) Auxin (b) Cytokininss
(c) Ethylene (d) Gibrelines

38. **Plant growth regulators normally produced in the leaves of plants, originating from chloroplasts, especially when plants are under stress?**
(a) Abscisic acid (AB(A) (b) Gibrelines (G(A)
(c) Cytokininss (CKs) (d) None of these

39. **Which one of the following is rare plant growth regulators having function as gravitropism, resistance to stress, and xylem differentiation?**
(a) Brassinosteroids (b) Jasmonates
(c) Plant peptide hormones (d) Salicylic acid

40. **For embryogenesis to germination, which one of the following hormone that permits the embryo to pass directly while still present on the parent plant?**
(a) ABA (b) Cytokinin (c) GA_3 (d) IAA

41. **In hybrid seed production programme, GA_3 plant hormone is sprayed on rice for which one of following?**
(a) Achieving nicking
(b) Enchaining exertion of inflorescence
(c) Enhancing time of stigma receptivity
(d) Inducing mal sterility

42. **Which one of the following is/are the most common auxin found in plants?**
(a) 2,4-D (b) 1-Naphthaleneacetic acid (NA(A)
(c) Indole-3-acetic acid (IA(A) (d) None of these

43. The physiological effects overcoming the need for vernalization in long-day plants, have been recorded by the application of which one of the following harmones?
(a) Abscics acids (b) Auxin
(c) Cytokinins (d) Gibrelines

44. Which of the following is precursor for synthesis of all natural cytokinins?
(a) Tryptophan *(b) S*-Adenosylmethionine
(c) [9R-5'-P] iP (d) None of these

45. Which one of the following is the major Auxin in developing seed?
(a) ABA (b) IAA
(c) NAA (d) Zeatin

46. Which one of the following is a volatile plant hormone?
(a) Auxine (b) Cytokinins
(c) Ethylene (d) Gibrelines

47. Dormin is the name of which one of the following phytohormones?
(a) Auxins (b) Cytokininss (CKs)
(c) Abscisic acid (AB(A) (d) Gibrelines (G(A)

48. Which one of the following is true statement about by brassinosteroids?
(a) Necessary for pollen elongation for pollen tube formation.
(b) Involvement in expression of many light regulated genes is exhibited.
(c) Provide some protection to plants during chilling and drought stress.
(d) All of the above

49. In the plants, which one of the following is/primery physiological effect of phytohormone auxin?
(a) Cell division (b) Cell elongation of in shoots
(c) Internodes elongation (d) All of the above

50. Transportation of gibberellin in the plants are ________.
(a) Lateral (b) Polar
(c) Non-polar (d) None of these

51. The first naturally occurring cytokinin was chemically identified in which one of the following years from maize (zea mays) grains and was called zeatin.
(a) 1962 (b) 1963
(c) 1964 (d) 1965

52. For hybrid seed production of rice, which one of following plant hormone requires?
(a) GA_3 (b) GA_6 (c) IAA (d) None of these

53. Which one of the following is true for ethylenes?
(a) Acceleration of sencsences (b) Delay of sencsences
(c) Inhibition of sencsences (d) Prevention of sencsences

54. If the leaf of plant is injured, it produces one of the following harmones which may cause the part to age and drop off.
(a) Abscisic acid (b) Auxins
(c) Cytokinins (d) Ethylene

55. Polyamines are an organic compound having two or more primary amino groups $-NH_2$ and are known as which one of the following?
(a) Polyvalent anions (b) Polyvalent cations
(c) Zwitter-ions (d) None of these

56. In germinating cereals, which one of the following phytohormone is responsible for conversion of stored proteins into glucose?
(a) Abscissic Acid (b) Auxin
(c) Cytokinin (d) Gibberellin

57. Gibberellins are found in different forms, which one of the following forms of Gibberellins refer to gibbrellicacids, a dihydroxylated gibberellin and first structurally characterized?
(a) GA_1 (b) GA_3 (c) GA_4 (d) GA_7

58. Which one of the following hormones have a stimulatory effect on the division of plant cells and retarding effect on leaf senescence?
(a) Auxin (b) Cytokininss
(c) Gibrelines (d) Ethylene

59. Which of the following is precursor in biosynthesis of ethylene?
(a) 1-Aminocyclopropane-1-carboxylic acid (AC(C)
(b) S-Adenosylmethionine (S-AdoMet)
(c) Both (a) and (b) (d) None of these

60. Abscisic acids (ABAs) are known as derivatives of which one of the following that synthesize in the MEP (plastidal 2-C-methyl-D-erythritol-4-phosphate) pathway and have different functions in many plant developmental processes, including bud dormancy?
(a) Momoterpenoid (b) Diterpenoid
(c) Triterpenoid (d) Sesqieterpenoid

61. In tissue culture media at concentrations of 10-1000 mM, which one of the following phytohormones are used that occur in high levels within plants?
(a) Auxins (b) Cytokinins
(c) Polyamines (d) None of these

62. Which one of the following concerned with gibberellin phytohormones?
(a) Bolting phenomenon is associated with gibberellin phytohormones.
(b) Gibberellin phytohormone is responsible for overcoming genetic dwarfism.
(c) Both (a) and (b) (d) None of these

63. First discovered hormone from the following is?
(a) Auxin (b) Cytokininss
(c) Ethylene (d) Gibrelines

64. All gibberellins are known as derivatives of which one of the following that synthesize by the terpenoid pathway in plastids and then modified in the endoplasmic reticulum and cytosol?
(a) Momoterpenoid (b) Diterpenoid
(c) Triterpenoid (d) Sesqieterpenoid

65. Kinetin is a type of cytokinin that promotes cell division and exhibits cytokinin activity in plants. This phytoharmone is originally isolated by which one of the following persons?
(a) Letham *et al.* (b) Miller *et al.*
(c) Zachau *et al.* (d) None of these

66. Which one of the following is considered with ethylene?
(a) Copper involves as cofactor in binding of ethylene to its receptor.
(b) Ethylene acceptors are located on endoplasmic reticulum (ER).
(c) Ethylene is known as fruit ripening phytohormones.
(d) All of the above

67. Which one of the following biosynthesis is required for seed maturation and dormancy during seed development?
(a) ABA biosynthesis (b) Auxine biosynthesis
(c) GA biosynthesis (d) None of these

68. The process of mitosis and meiosis affected by which one of the following is rare plant growth have been found in all organisms studied thus far?
(a) Nitric oxide (b) Plant peptide hormones
(c) Polyamines (d) None of these

69. Which one of the following is a standard bioassay method for cytokinin?
(a) Carrot root phloem bioassary (b) Cell-division test
(c) Chlorophyll preservation test (d) Cotyledon expansion test

70. In water stress, which one of the following phytohormone play an important role?
(a) Auxin (b) Absiciic acid
(c) Ethylene (d) Gibberlic acid

71. Gibberellins are tetracyclic diterpene acids. They are distinguished by containing carbon atom in their structure is?
(a) 19 carbons (b) 20 carbons
(c) 19 carbons or 20 carbons (d) None of these

72. GA_3 is applied in hybrid seed production of which among the following crops?
(a) Maize (b) Pearl millet
(c) Rice (d) All of the above

73. Which one of the following is correct name of name of kinetin?
(a) 5-furfurylaminopurine (b) 5-furfurylaminopyrimidine
(c) 6-furfurylaminopurine (d) 6-furfurylaminopyrimidine

74. A plant to drop fruits and leaves caused by increase in ______ amount and decrease in ______ amount in fruit and leaf stalks.
(a) Auxin and Ethylene (b) Cytokinin and Abscisic acid
(c) Ethylene and Auxin (d) Gibberellin and Abscisic acid

75. Prevention of sencsences is a physiological role of ________ plant hormones.
(a) Cytokinines (b) Ethylinens
(c) Jasminic acids (d) None of these

76. Which one of the following test is a bioassay method for gibberellins, except?
(a) Avena coleoptile test (b) Barley endosperm test
(c) Dwarft corn test (d) Lettuce hypocotyl test

77. Auxin is synthesized in which part of the plants?
(a) Root (b) Shoot
(c) Meristermatic regions (d) Both (a) and (b)

78. **Till now, there were more than 126 GAs have been identified from which ome of the following sources?**
(a) Fungi and Bacteria (b) Plants and Bacteria
(c) Plants and Fungi (d) Plants, Fungi and Bacteria

79. **Which one of the following are true statements?**
(a) Delay of senescence is physiological effects specific to kinetin.
(b) Cytokinin has not been isolated from plants.
(c) Trans-zeatin is biologically most active natural cytokinin in plants.
(d) All of the above

80. **In plants, naturally occurring growth inhibitor is which of the following?**
(a) Abscisic acid (AB(A)
(b) Gibrelic acid-41 (GA_1)
(c) Indole-3-acetic acid (IA(A)
(d) All of the above

81. **Which one of the following plant growth regulators are produced from fatty acids to promote the production of defense proteins that are used to fend off invading organisms?**
(a) Brassinosteroids (b) Jasmonates
(c) Salicylic acid (d) None of these

82. **Which one of the following hormone should be present at low concentration for germination of seed?**
(a) ABA (b) Auxin
(c) Cytokinin (d) Gibberlin

83. **Which one of the following is a correct statement?**
(a) 2.4-D, TIBA, DCA are known as antagonist auxin.
(b) Abscisic acid (AB(A) controls leaf fall in plants and dormancy in seeds.
(c) Bioassay implies using living organisms to test the biologically active chemicals.
(d) Parthenocarpies in fruits are produced by spraying auxins on flowers.

84. **In 1931, an active substance isolated from human urine by Kogl and Haagen Smit which was named as one of the following?**
(a) Auxin-A (b) Auxin-B
(c) Auxen triolic acid (d) Both (a) and (c)

85. In 1926, gibberellin was first recognized which one of the following scientist by studying of bakanae, the "foolish seedling" disease in rice?
(a) Eiichi Kurosawa (b) Teijiro Yabuta and Hayashi
(c) Teijiro Yabuta and Sumuki (d) None of these

86. Which one of the following is true statements, except?
(a) Abscisic acid (AB(A) phytohormones regulate to stomatal closure.
(b) Cytokinins phytohormone shows Richmond-Lang effect.
(c) Ethylenes are regarded as a true hormone.
(d) Kinetin was first isolated from DNA of herring sperm.

87. Physiological role of abscisic acid (AB(A) is which of the following?
(a) Delay of senescence (b) Dormancy of seeds and buds
(c) Fruit ripening and senescence
(d) Both (b) and (c)

88. Which one of the following plant growth regulators are not produced from fatty acids and not protein derivatives but activates genes in some plants that produce chemicals that aid in the defense against pathogenic invaders?
(a) Brassinosteroids (b) Jasmonates
(c) Plant peptide hormones (d) Salicylic acid

89. Kogl and Haagen Smit (1934) isolated an active substance from malt and corn-grain oil which was called as one of the following?
(a) Auxin-A (b) Auxin-B
(c) Auxen triolic acid (d) Both (a) and (c)

90. Which one of the following scientists named as gibberellin and first isolated in 1935 from fungal strains *Gibberella fujikuroi* (perfect state of *Fuasarium moniliforme*) and named also?
(a) Eiichi Kurosawa (b) Teijiro Yabuta and Hayashi
(c) Teijiro Yabuta and Sumuki (d) None of these

91. Abscisic acid (AB(A) involved in stomata closure in which one of the following?
(a) Heat stressed plants. (b) Salt stressed plants.
(c) Water stressed plants. (d) None of these

92. Plant growth regulators are involved in cell-to-cell signaling; and also play crucial roles in pollen self-incompatibility?
(a) Nitric oxide (b) Plant peptide hormones
(c) Polyamines (d) None of these

93. In 1934, who one of the following was first pointed that the apical dominance under control of auxin produced at the terminal bud and transported downword through the stem.
(a) M. Bayliss and E.H. Starling (b) Scoog and Thimann
(c) Teijiro Yabuta and Sumuki (d) None of these

94. Which one of the following is the primary precursor for the synthesis of gibberelines?
(a) Acetic acid (b) Oxaloacetic acid
(c) Pyruvic acid (d) Both a0 and (b)

95. Which of the following phytohormones promoting to the fruit ripening?
(a) Abscisic acid (b) Ethylene
(c) Gibberellin (d) Both (a) and (b)

96. Can be regulate plant cell organelle functions e.g. ATP synthesis in chloroplasts and mitochondria. Which one of the following is PGRs play this role?
(a) Nitric oxide (b) Plant peptide hormones
(c) Polyamines (d) None of these

97. Which one of the following is true statement about auxins in plant, except?
(a) Conjugated or bound auxins complexes with carbohydrates.
(b) Conjugated or bound auxins complexes with inositol.
(c) Conjugated or bound auxins complexes with lipids.
(d) Conjugated or bound auxins complexes with proteins or amino.

98. Which one of the following amino acids is precursor of auxin in plants?
(a) Proline (b) Tryptophan
(c) Tyrosine (d) Both (b) and (c)

99. In higher plants, which of the following is true for physiological effect of auxin, except?
(a) Auxin enhance to the apical dominance.
(b) Auxin prevents to senescence in plants.
(c) Auxin stimulates to cell elongation in plants.
(d) Auxin high conc. initiats more lateral bramch roots.

100. In the plants, which one of the following is the result of auxin's primary physiological effect?
(a) Cell division (b) Elongation of cells in shoot
(c) Elongation of internodes (d) All of the above

101. Transportation of auxin in the plants is ________.
(a) Lateral (b) Polar
(c) Non-polar (d) None of these

102. In which one of the following parts of plant, gibberellins are stores?
(a) Immature seeds (b) Leaf
(c) Root (d) Stem

103. Which one of the following are synthetic auxins that stimulate uncontrolled DNA and protein synthesis and are effective herbicides?
(a) á-Naphthalene acetic acid(á-NA(A)
(b) 2,4,5-trichlorophenoxyacetic acid (2,4,5-T)
(c) 2,4-dichlorophenoxy acetic acid (2,4-(D)
(d) Both (b) and (c)

104. Auxin-cytokinin ratio was first recognized in the 1950s by which one of the following scientist that refers to the relative proportion of auxin to cytokinin present in plant-tissue-culture media?
(a) C. O. Miller and F. Skoog (b) M. Bayliss and E.H. Starling
(c) Teijiro Yabuta and Sumuki (d) None of these

105. The specific physiological effect of gibberellins is, except?
(a) Elongation of intemodes (b) Bolting and flowering
(c) *de novo* synthesis of á-amylase (d) Prevention of abscission

106. At cell division and elongation in embryo and endosperm, which one of the following hormone remains present at its highest concentration?
(a) ABA (b) Auxin
(c) Cytokinin (d) Gibberlin

107. Which one of the following is considered with auxin?
(a) Auxin is degraded chiefly by oxidation by O_2 in in presence of IAA oxidase.
(b) F. W. Went developed 'avena-curvature test' for bioassay of auxin.
(c) Relative auxin conc. on two sides of abscission zone, control to abscission of leaves
(d) All of the above

108. F. W. Went discovered which one of the following test to find out the amount of growth promoting hormones?
(a) Avena coleoptiles test (b) Carrot root phloem bioassary
(c) Lettuce hypocotyl test (d) None of these

109. Which of the following naturally occurring (endogenous) auxins were found to be synthesized by plants?
(a) Phenaylacetic acid and Indole-3-butyric acid
(b) 1-Naphthaleneacetic acid and 2,4-dichlorophenoxy acetic acid
(c) Indole-3-acetic acid and 4-chloroindole-3-acetic acid
(d) Both (a) and (c)

110. Which one of the following are true statements?
(a) Cycocel (CCC) is growth retardant, use in floriculture industry.
(b) Naturally occurring, biologically active ABA in plants exists in *Cis* form.
(c) Trans-zeatin is biologically most active natural cytokinin in plants.
(d) All of the above

111. GA-induced which one of the following pathways are Ca^{+2}-independent, leads to á–amylase expression in cells of aleurone layer?
(a) Glycolytic pathway (b) Signal transduction pathway
(c) â-oxidation pathway (d) None of these

112. The physiological effect of auxin in plants is which one of the following?
(a) Apical dominance (b) Elongation of the cells
(c) Inition of roots (d) All of the above

113. Which one of the following is a chemical/hormone which inhibits and promote to the seed germination?
(a) Cytokinin (b) Ethylene
(c) Gibberellin (d) None of these

114. Which of the following phytohormones is sesquiterpene?
(a) Abscisic acid (b) Ethylene
(c) Gibberellin (d) None of these

115. Which of the following hormone's biosynthesis is required for seed maturation and dormancy during seed development?
(a) Abscisic acid (b) Auxins
(c) Cytokinins (d) Gibberellins

116. Which of the following combines with abscisic acid and ethylene to protect the plant from predation and pathogen attack?
(a) Jasmonic acid (b) Brassinosteroids
(c) Salicylic acid (d) None of these

117. Which of the following hormones does not naturally occur in plants?
(a) 2, 4 – D (b) Abscisic acid (AB(A)
(c) Gibberellin (G(A) (d) Indole-3-acetic acid (IA(A)

118. What will be shape of the curve, if plant growth rate is plotted against time?
(a) N-shaped (b) S-shaped

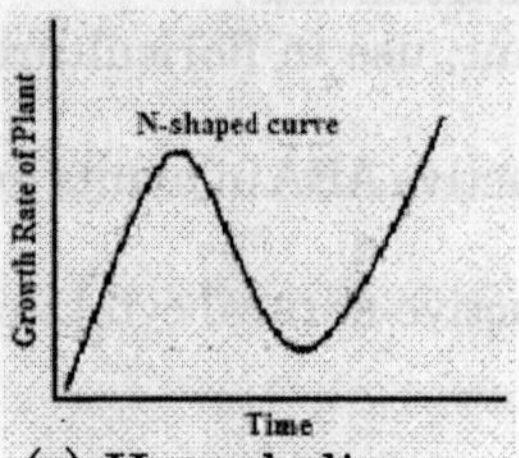

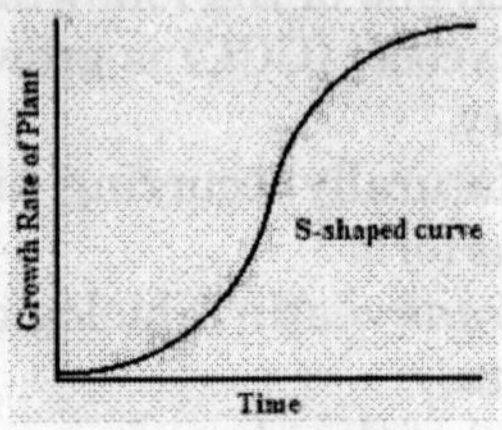

(c) Hyperbolic (d) None of these

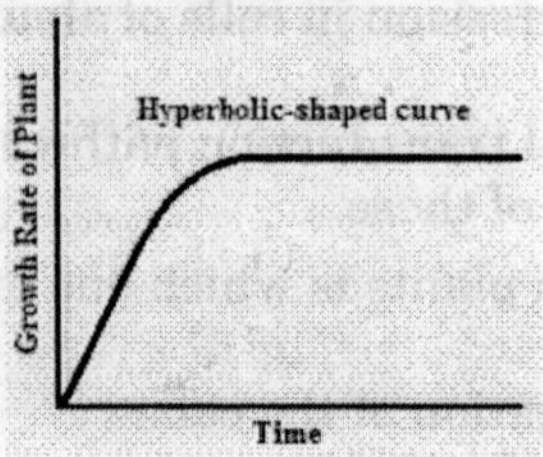

ANSWERSHEET

8. Plant Hormones

1. b	2. b	3. b	4. c	5. b	6. d	7. c
8. b	9. d	10. b	11. b	12. c	13. d	14. c
15. b	16. d	17. a	18. c	19. b	20. c	21. c
22. b	23. a	24. d	25. c	26. d	27. d	28. a
29. c	30. b	31. a	32. d	33. c	34. d	35. b
36. a	37. a	38. a	39. c	40. d	41. b	42. c
43. d	44. c	45. b	46. c	47. c	48. d	49. b
50. c	51. b	52. a	53. a	54. a	55. b	56. b
57. b	58. b	59. c	60. d	61. c	62. c	63. a
64. b	65. b	66. d	67. a	68. c	69. a	70. a
71. c	72. c	73. c	74. d	75. c	76. a	77. c
78. d	79. d	80. a	81. b	82. a	83. a	84. d
85. a	86. c	87. d	88. d	89. d	90. c	91. c
92. b	93. b	94. a	95. d	96. a	97. c	98. b
99. b	100. b	101. b	102. a	103. d	104. a	105. d
106. c	107. d	108. a	109. d	110. d	111. b	112. d
113. b	114. a	115. a	116. a	117. a	118. b	

9

Gene, Gene Expression and Regulation

1. W. Johansen in 1901, coined which one of the following term?
   (a) Gene (b) Genetic code
   (c) Genome (d) Transcription
2. Initiation of protein synthesis requires which one of the following is/are?
   (a) Initiation factor (b) mRNA
   (c) Ribosome subunit (d) All the above
3. Which one of the following scientist showed that genes reside on specific chromosomes later explained genes occupy specific locations on the chromosome?
   (a) Charles Darwin (b) Thomas Hunt Morgan
   (c) Wilhelm Hofmeister (d) Wilhelm Johannsen
4. Which one of the following scientist showed by an experiment that genes could be transferred?
   (a) Frederick Griffith (b) Thomas Hunt Morgan
   (c) Wilhelm Hofmeister (d) Wilhelm Johannsen
5. Protein is a polymer of which one of the following?
   (a) Amino acid (b) Carbohydrate
   (c) Fat (d) All the above
6. Which one of the following is a process of synthesis of mRNA from a DNA template?
   (a) Transcription (b) Transduction
   (c) Transformation (d) Translation
7. Which one of the following is a process of synthesis of DNA from RNA?
   (a) Reverse mutation (b) Reverse Transcription
   (c) Reverse translation (d) Transcription

8. **Synthesis of mRNA only from one strand of DNA is known as which one of the following?**
(a) Asymmetrical transcription (b) Symmetrical transcription
(c) Teminism (d) None of these

9. **Reverse transcription has been reported in which one of the following?**
(a) *Drosophila* (b) *E.Coli*
(c) *Neurospora* (d) Some tuber producing viruses

10. **The evidence for intragenic recombination was presented by which one of the following in 1940.**
(a) Benzer (b) McClincton
(c) Morgan (d) Oliver

11. **Which one of the following concerned with cistron is/are?**
(a) A section of the DNA or RNA molecule.
(b) Functional unit of the hereditary materials.
(c) Specifies the formation of one polypeptide chain.
(d) All of the above

12. **For which one of the following the discovery, H. M. Temin and David Baltimore were awarded Nobel Prize?**
(a) Reverse Mutation (b) Reverse Transcription
(c) Transcription (d) Translation

13. **Cistron is derived from cis and trans, the term cistron was coined in 1957 by which one of the following?**
(a) Baltimore (b) Beadle
(c) Benzer (d) Leaderberg

14. **Which one of the following is codes for a specific gene product either a protein or RNA?**
(a) Cistron (b) Operon
(c) Both (a) and (b) (d) None of these

15. **Which one of the following is the main function of a *cis-acting* gene?**
(a) Encodes a protein (b) Encodes an RNA only
(c) Encodes an RNA or a protein (d) Protein binding

16. **Structural genes serve the function of which one among the following is/are?**
(a) Binding site for protein (b) Encodes an RNA
(c) Both (a) and (b) (d) None of these

17. **Which of the following functions is performed by members of a gene family?**
(a) Different (b) Identical (c) Related (d) None of these

18. In eukaryotes, gene occurs in gene cluster. Which of the following is/are?
(a) Histone (b) snRNA
(c) tRNA (d) All of the above

19. The biosynthetic pathway of which of the following is/are, where feedback inhibition is common?
(a) Amino acids (b) Purines and Pyrimidines
(c) Vitamins (d) All of the above

20. A specific stimulus is needed for the induction of synthesis of specific enzymes. The first evidence was discovered by which one of the following?
(a) F. Dienert (b) F. Jacob
(c) J. Monad (d) None of these

21. β-Galactosidase enzyme splits lactose into which one of the following?
(a) Fructose and Galactose (b) Glucose and Fructose
(c) Glucose and Galactose (d) None of these

22. For which of the following enzymes, lactose acts as inducer?
(a) Galactoside permiase (b) Thiagalactoside acetylase
(c) β-galactosidase (d) All of the above

23. Which one of the following theories illustrates how certain genes can be active or inactive?
(a) The germplasm theory (b) The operon hypothesis
(c) Theory of acquired character (d) None of these

24. Which one of the following is correct?
(a) Introns are present in mRNA and exons are present in tRNA.
(b) Codons are present in mRNA and anticodons in tRNA.
(c) Every intron is a set of three terminator codons.
(d) Exons are present in eukaryotes while introns are present in prokaryotes.

25. Operons are as a cluster of genes under the control of a single regulatory signal or promoter. Which one of the following were developed concepts of the operon?
(a) E. Englesberg and G. Wilcox (b) E. W. Johns and T.A. Hoare
(c) F. Jacob and J. Monad (d) J. R. Beckwith and D. Zisper

26. Genes which code for more than one protein are known which one of the following genes?
(a) Overlapping genes (b) Pseudo genes
(c) Selfish genes (d) Split genes

27. Which one of the following genes refers to defective copies of normal genes?
(a) Overlapping genes (b) Pseudo genes
(c) Selfish genes (d) Split genes

28. With no obvious function, a repetitive DNA sequence is known as which one of the following genes?
(a) Overlapping genes (b) Pseudo genes
(c) Selfish genes (d) Split genes

29. Which one of the following concerned with the operon is/are?
(a) Control of the synthesis of a protein or a group of proteins.
(b) Under the control of an operator gene
(c) Unit of coordinated and regulated gene activity found in prokaryotes.
(d) All of the above

30. Which one of the following genes determines the organization of the proteins is controlled by other regions of DNA?
(a) Regulator gene (b) Structural genes
(c) None of these (d) Both (a) and (b)

31. The operator and its associated genes are called as an operon. The operon consists of a segment of genomic DNA containing a ________________ or a linear sequence of ________________.
(a) Regulator gene (b) Structural genes
(c) Both (a) and (b) (d) None of these

32. "A model of genetic regulation of protein synthesis in prokaryotes whereby the structural genes determines to the proteins organization, are controlled by other DNA regions upstream from the structural genes". Which one of the following models is?
(a) Britten-Davidson model
(b) Jacob-Monod model
(c) Koshland model (d) Monod-Wyman-Changeux model

33. Association of which one of the following, the operator DNA prevents the transcription of the operon in negative control of gene expression?
(a) Glycoprotein (b) Lipoprotein
(c) Protein (d) RNA

34. What will be the codons in m-RNA if the DNA codes are ATG-CAG?
(a) TAC – GTC (b) TCA – GTC
(c) UAC – GUC (d) UCA – TUA

35. Color blindness is governed by a recessive gene located on which one of the following?
(a) Autosomes (b) X chromosome
(c) Y chromosome (d) None of these

36. The holandric gene is linked with which one of the following chromosome?
(a) X-chromosome (b) Y-chromosome
(c) Both (a) and (b) (d) None of these

37. Which one of the following is true about holandric genes?
(a) Genes carried in Y chromosomes and thereby showing only male to male transmission.
(b) Genes responsible for determination of sex in individual.
(c) The sex having genes in heterozygous conditions.
(d) All of the above

38. In *E. coli*, operons are known for metabolism of which one of the following is/are?
(a) Arabinose (b) Arginine
(c) Tryptophan (d) All of the above

39. Biosynthetic pathways show repressible negative control in the operon. Which one of the following is/are responsible for?
(a) Biosynthesis of Histidines
(b) Biosynthesis of Purines/Pyramidines
(c) Biosynthesis of Tryptophan (d) All of the above

40. Which one of the following serve as translation initiator codon, in case of leader sequence of the *try* operon of *E. coli*?
(a) AGU (b) AUG
(c) GUA (d) GUG

41. Name the scientist who developed eukaryotes "operon" model?
(a) Britten and Davidson
(b) E. Englesberg and G. Wilcox
(c) E. W. Johns and T. A. Hoare
(d) J. R. Beckwith and D. Zisper

42. Gene action may be regulated at which one of the following levels in eukaryotes?
(a) After translation (b) Gene replication
(c) Transcription and Translation (d) All of the above

43. In the eukaryotic opron model, signal receiving gene is known as which one of the following?
(a) Integrator gene (b) Producer gene
(c) Receptor gene (d) Sensor gene

44. Which one of the following is a model that state "gene regulation in eukaryotes that is somewhat similar to the operon model (Jacob and Monod model) in prokaryotes?"
(a) Britten-Davidson model (b) Koshland model
(c) Monod-Wyman-Changeux model
(d) None of these

45. Which one of the following is a gene that initiates the transcription and located near the producer gene in eukaryotes?
(a) Integrator gene (b) Receptor gene
(c) Sensor gene (d) None of these

46. In Britten-Davidson model, postulates two levels of positive control, requiring two classes of regulatory proteins, the first control level consists of the interaction of a primary regulatory protein. Which one of the following is/are considered by this model?
(a) The second control level consists of the synthesis of secondary regulatory proteins.
(b) Activation of the sensor gene results in an activator RNA production.
(c) Activator proteins interact with receptor sites to promote structural gene expression.
(d) All of the above

47. Which one of the following is a gene responsible for the transcription and protein synthesis; located near the receptor gene in eukaryotes?
(a) Integrator gene (b) Producer gene
(c) Sensor gene (d) None of these

48. Which one of the following is a gene that receives signal from sensor gene and initiates the transcription in eukaryotes?
(a) Integrator gene (b) Producer gene
(c) Receptor gene (d) None of these

49. Which one of the following is/are considered with gene regulation in eukaryotes?
(a) Differential mRNA degradation
(b) Differential mRNA localization
(c) Ribosome modification (d) All of the above

50. The transcription activation domain of regulatory transcription factor may serve as which one of the following?
(a) Acidic domain (b) Glutmine-rich regions
(c) Proline-rich region (d) All of the above

51. In which one of the following, gene position effects have been reported?
(a) Drosophila and Neurospora (b) Maize and Mice
(c) Oenothera and Human (d) All of above

52. The existence of loci controlling elements has been demonstrated by which one of the following scientists?
(a) McCarty (b) McClintock
(c) Stadler (d) Stern

53. The term, 'southern blotting' refers to which one of the following?
(a) Attachment of probes to DNA fragments.
(b) Comparison of DNA fragments from two sources.
(c) Transfer of DNA fragments from electrophoresis gel to nitrocellulose sheet.
(d) Transfer of DNA fragments from *in vitro* cellulose membrane to electrophoresis gel.

54. In the RNA and DNA molecule which one of the following purine bases is found?
(a) Adenine (b) Cytocine
(c) Guanine (d) Thymine

55. Operons are found in which one of the following?
(a) Drosophila *(b) E. Coli*
(c) Yeast (d) Both (a) and (b)

56. An operon is made up of which one of the following?
(a) Operator (b) Promoter
(c) Structural genes (d) All of the above

57. Which one of the following is/are considered with monitoring gene expression?
(a) Amount of protein (b) Amount of RNA
(c) Intensity of phenotype (d) All of above

58. Which one of the following is where glucocorticoid response element found?
(a) Enhancer (b) Operator
(c) Promoter (d) Terminator

59. Which one of the following is/are correct statement considered with enhancers?
(a) Binding site for proteins.
(b) Organization is similar to that of promoters.
(c) Some binding sites are common with promoters.
(d) All of above

60. Generally at which of the following stages, gene expression is regulated?
(a) Post transcriptional modification
(b) Post translational modification
(c) RNA processing (d) Transcription

61. In the regulation of the expression of which one of the following gene level, gene amplification involves?
(a) mRNA gene (b) rRNA gene
(c) tRNA gene (d) All the above

62. Which one of the following autogenous repressions regulates to the production of ribosomal protein?
(a) mRNA destruction (b) Transcription
(c) Translation (d) All the above

63. Protein directly controls to the expression of its own structural gene through the mechanism is called as autogenous regulation. Autogenous repression is related to which one of the following?
(a) mRNA destruction (b) Transcription
(c) Translation (d) Both (b) and (c)

64. Tubulin is a dimer composed of a- and â- subunits and the principal protein component of microtubules. Which one of the following regulates to the production of tubulin proteins?
(a) Gene amplification (b) mRNA destruction
(c) RNA processing (d) Translation repression

65. Epigenetics is the analysis of the mechanisms causing phenotypic effects to be produced by the genes of a genotype. Which one of the following is/are example of epigenetic inheritance?
(a) Prions (b) Genetic Imprinting
(c) Permutation (d) All of the above

66. Which one of the following is/are true regarding prions?
(a) Formerly known as slow viruses.
(b) Devoid of nucleic acids.
(c) Neither viruses nor viroids. (d) All of the above

67. Which one of the following considered with genomic imprinting?
(a) An inheritance process independent of the classical Mendelian inheritance.
(b) In a parent of origin specific manner, genetic phenomenon expressed to the certain genes.
(c) Occurs in the ovary or testis early in the formation of the eggs and sperm.
(d) All of the above

68. Which one of the following is a true statement regarding genomic imprinting?
(a) An epigenetic process that involves DNA methylation and histone and methylation.
(b) Have been demonstrated in insects, mammals and flowering plants.
(c) Achieved monoallelic gene expression without altering the genetic sequence.
(d) All of the above

69. Permutation is a mutation phenomenon discovered in corn. Which one of the following is true statements regarding permutation?
(a) In the heterozygous condition, one allele permanently changes the partner allele.
(b) One allele (first) refers to paramutagenic and another allele (secon(d) to paramutable.
(c) Both (a) and (b) (d) None of these

70. Which one of the following concerned with feedback inhibition is?
(a) Biosynthetic pathways (b) Enzyme activity
(c) Both (a) and (b) (d) None of these

71. Regular hetero-hromatinization leading to dosage compensation occurs in
(a) Drosophila (b) Humans (c) Mice (d) Both (b) and (c)

72. A regulator gene mutation producing a repressor having defective DNA binding site will become which of the following in case of operon under inducible negative control?
(a) Constitutive (b) Super Repressed
(c) Un-inducible (d) None of these

73. What happens if the regulator protein had a defective site for binding of the co-repressor and an operon under repressible negative control?
(a) Easily Repressed (b) Non Repressible
(c) Super Repressed (d) Uninducible

74. About how many inducer molecules is binding by an active *E. coli* lac repressor?
(a) 1 (b) 2 (c) 3 (d) 4

75. Split gene was first reported in which one of the following?
(a) Chickens *(b) Drosophila*
(c) Garden Pea (d) None of these

76. **Genes with intervening sequences are called as which one of the following?**
(a) Jumping genes (b) Overlapping genes
(c) Pseudo genes (d) Split genes

77. **The terms exons and introns are related to which one of the following?**
(a) Jumping genes (b) Overlapping genes
(c) Pseudo genes (d) Split genes

78. **Coding sequences in split genes are called which one of the following?**
(a) Exons (b) Introns (c) Mutons (d) Recons

79. **Non-coding sequences In split gene are known as which one of the following?**
(a) Exons (b) Cristrons (c) Introns (d) Mutons

80. **The largest unit within a gene is which one of the following?**
(a) Cistron (b) Intron (c) Muton (d) Recon

81. **The smallest unit within the gene is which one of the following?**
(a) Cistron (b) Intron (c) Muton (d) Recon

82. **Which one of the following is termed as the unit of function within the gene?**
(a) Cistron (b) Intron (c) Muton (d) Recon

83. **The unit of recombination within a gene is termed as which one of the following?**
(a) Cistron (b) Intron (c) Muton (d) Recon

84. **Which one of the following is the unit of mutation within a gene?**
(a) Cistron (b) Intron (c) Muton (d) Recon

85. **An *E.coli* repressor to relieve repression has must binded a minimum of how many inducer molecules?**
(a) 1 (b) 2 (c) 3 (d) 4

86. **Which one of the following features is all CAP (catabolite activator protein) dependent promoters?**
(a) Poorly Conserved -10 Sequence
(b) Poorly Conserved -35 Sequence
(c) Well Conserved -10 Sequence
(d) Well Conserved -35 Sequence

87. **Which one of the following is used Anti-termination is?**
(a) Some Bacterial Operon (b) Some Phage Gene
(c) Both (a) and (b) (d) None of these

88. Which one of the following affects to the anti-termination is?
(a) RNA polymerase core enzyme
(b) RNA polymerase holoenzyme
(c) Terminator (d) σ factors

89. In *E. coli*, σ factors influence RNA polymerase to specifically transcribe specific sets of genes most likely by interacting with which one of the following?
(a) -10 Sequence (b) -35 Sequence
(c) Both (a) and b (d) None of these

90. Over 20 dwarfing genes have been identified; the most commonly used in wheat breeding is *Rht1*, *Rht2*, and *Rht8*. Which of the following is designated as Norin 10 dwarfing genes?
(a) *Rht1* (b) *Rht2* (c) *Rht8* (d) Both (a) and (b)

91. A specific σ factors regulates to the transcription of which one of the following?
(a) Alternative nitrogen source utilization gene
(b) Flageller Genes
(c) Heat and Shock Gene (d) All of the above

92. The gene causing Hemophilia in man is inherited through which one of the following?
(a) Males (b) Females
(c) Both (a) and (b) (d) None of these

93. Sex linked genes are located on which one of the following?
(a) X-chromosomes (b) Y-chromosomes
(c) Autosomes (d) None of these

94. For paramutation, explanation seems to be most correct is among the following?
(a) A subtle but critical change in gene sequence by gene conservation
(b) Change in chromatin structure
(c) Increased number of copies of a heterochromatic segment.
(d) Methylation at C-residue.

95. Dosage compensation by hyperactivity of X- linked gene is found in which one of the following?
(a) Birds (b) *Drosophila*
(c) Mice (d) Reptiles

96. The expression of different sets of structural gene, have been affected by different controlling elements in case of maize. The smallest number of structural gene is affected by which one of the following?
(a) *Ac-Ds* (b) *Dt* (c) *spm-dspm* (d) Both (b) and (c)

97. Regulatory mechanisms of which one of the following dependent gene expression have been studied in maize using a vp1 (viviparous1) mutant?
(a) Abscisic acid (AB(A) (b) Gibbereline (G(A)
(c) Cytokininss (CKs) (d) Ethylene

98. Which of the following is considered with *Ds* element of maize?
(a) It cannot transpose on its own and is stable.
(b) It is derived from *Ac*.
(c) It is generally deleted version of *Ac*. (d) All of the above

99. The expression of different sets of structural gene, have been affected by different controlling elements in case of maize. The largest number of structural gene is affected by which one of the following?
(a) *Ac-Ds* (b) *Dt* (c) *spm-dspm* (d) Both (a) and (c)

100. The allele contributed by which one of the following parents is functional in case of genetic imprinting in mammals?
(a) Male (b) Female
(c) Either male or female (d) Both male and female

101. Which of the following is the predominant mode of gene regulation in eukaryotes?
(a) Negative (b) Positive
(c) Combined Positive-Negative (d) None of these

102. Genetic imprinting involves in which one of the following is/are?
(a) Enzyme Action (b) Regulatory Protein
(c) Structural Protein (d) All of the above

103. The DNA binding domain of a eukaryotic regulatory factor is/are which one of the following?
(a) Amphipathic helix-loop-helix. (b) Glutamine-rich region
(c) Leucine zipper (d) Both (a) and (c)

104. As an anti- terminator factor, which one of the following protein functions is/are?
(a) E.coli protein NusB-NusE (b) Phage ë protein pN
(c) Phage ë protein pQ (d) All of the above

105. In transcription of most of *E. coli* genes, which one of the following ó-factors are involved?
(a) σ^{28} (b) σ^{32} (c) σ^{54} (d) σ^{70}

106. in transcription of heat shock genes, which one of the following σ factors are involved?
(a) σ^{28} (b) σ^{32} (c) σ^{54} (d) σ^{70}

107. Jumping gene was first discovered in which one of the following?
(a) *Drosophila* (b) Maize
(c) *Neurospora* (d) *Oenothera*

108. The proteins which code for a pair of contrasting traits are known as which one of the following?
(a) Alleles (b) Genes
(c) Both (a) and (b) (d) None of these

109. The genes are subject to Mendelian inheritance which are transmitted according to the laws of Mendel through the processes of nuclear division is?
(a) Integrator gene (b) Nuclear genes
(c) Regulator gene (d) Structural genes

110. Gene expression is results of a combination, except which one of the following?
(a) Altered protein activity
(b) Enhance translation
(c) Decrease amounts of specific mRNA
(d) Stabilize proteins

ANSWERSHEET

9. Gene, Gene Expression and Regulation

1. a	2. d	3. b	4. a	5. a	6. a	7. b
8. a	9. d	10. d	11. d	12. b	13. c	14. a
15. d	16. c	17. b	18. d	19. d	20. a	21. c
22. d	23. b	24. b	25. c	26. a	27. b	28. c
29. d	30. b	31. b	32. b	33. c	34. c	35. b
36. b	37. a	38. d	39. d	40. b	41. a	42. d
43. d	44. a	45. b	46. d	47. b	48. a	49. d
50. b	51. d	52. b	53. c	54. a	55. d	56. d
57. a	58. a	59. d	60. d	61. b	62. c	63. d
64. b	65. d	66. d	67. d	68. d	69. c	70. c
71. d	72. a	73. b	74. d	75. a	76. d	77. d
78. a	79. c	80. d	81. d	82. a	83. d	84. c
85. b	86. b	87. c	88. a	89. a	90. d	91. d
92. c	93. a	94. d	95. b	96. b	97. a	98. d
99. a	100. c	101. b	102. a	103. d	104. d	105. d
106. b	107. b	108. c	109. b	110. c		

10

Genome and Genomic Organizations

1. What is the genome size of Rice (*Oryza sativa* spp. *javanical*)?
   (a) 145 Millions of Base-Pairs (Mbp)
   (b) 262 Millions of Base-Pairs (Mbp)
   (c) 367 Millions of Base-Pairs (Mbp)
   (d) 424 Millions of Base-Pairs (Mbp)
2. What is the genome size of Arabidopsis (*Arabidopsis thalian*(a)?
   (a) 145 Millions of Base-Pairs (Mbp)
   (b) 262 Millions of Base-Pairs (Mbp)
   (c) 367 Millions of Base-Pairs (Mbp)
   (d) 424 Millions of Base-Pairs (Mbp)
3. Which one of the following methods has been used to estimate genome size?
   (a) DNA-reassociation kinetics.
   (b) Nuclear volume measurements.
   (c) Micro densitometry of Feulgen-stained nuclei.
   (d) All of the above
4. The first bacterial genome to be sequenced was which one of the following?
   (a) *Arabidopsis thaliana* (b) Bacteriophage MS2
   (c) Haemophilus *influenzae* (d) Phage Φ-X174
5. Which one of the following was first sequenced DNA-genome in virus organism?
   (a) *Arabidopsis thaliana* (b) Bacteriophage MS2
   (c) Haemophilus *influenzae* (d) Phage Φ-X174
6. Which one of the following was first sequenced RNA-genome in virus organism?
   (a) *Arabidopsis thaliana* (b) Bacteriophage MS2
   (c) Haemophilus *influenzae* (d) Phage Φ-X174

7. Which one of the following was first plant genome sequenced?
(a) *Arabidopsis thaliana* (b) Bacteriophage MS2
(c) Haemophilus *influenzae* (d) Phage Φ-X174

8. Phage Φ-X174 was first sequenced DNA-genome in virus, established which one of the following scientist?
(a) Fred Sanger (b) Hans Winkler
(c) Walter Fiers (d) None of these

9. Bacteriophage MS2 was first sequenced RNA-genome in virus, established which one of the following scientist?
(a) Fred Sanger (b) Hans Winkler
(c) Walter Fiers (d) None of these

10. Study of the physical nature of genomes and to determine and analyze the DNA sequence of the genome as its primary goal is known as which one of the following?
(a) Comparative genomics (b) Functional genomics
(c) Structural genomics (d) None of these

11. Study concerned with the genome functions. That is, it examines the transcripts produced by the genome and the array of proteins they encode.
(a) Comparative genomics (b) Functional genomics
(c) Structural genomics (d) None of these

12. Study concerned with the genomes from different organisms to look for significant differences and similarities is known as which one of the following?
(a) Comparative genomics (b) Functional genomics
(c) Structural genomics (d) None of these

13. The genetic materials play a fundamental role in determining the structure and nature of cell, found in which one of the following cell organelles?
(a) Nucleus and Cytoplasm (b) Nucleus and Mitochondria
(c) Nucleus, Mitochondria and Cytoplasm
(d) None of these

14. When DNA replication starts as which one of the following?
(a) The bonds between the nitrogen base and deoxyribose sugar break.
(b) The hydrogen bonds between the nucleotides of two strands break.
(c) The leading strand produces okazaki fragments.
(d) The phosphodiester bonds between the adjacent nucleotides break.

15. The genetic material of a cell can be a part of which one of the following of an organism?
(a) Gene (b) DNA molecule
(c) Genome (d) All of the above

16. Genetic material consists of which one of the following?
(a) DNA (b) RNA
(c) DNA and RNA (d) None of these

17. Maximum length of chromosomes is observed during which one of the following phase?
(a) Anaphase (b) Interphase
(c) Metaphase (d) Prophase

18. An organism or cell with two chromosome sets (2n) or two genomes in Somatic tissues of higher an individual with two chromosome sets in Somatic tissues of higher plants are known as which one of the following?
(a) Diploid (b) Disomic
(c) Both (a) and (b) (d) None of these

19. Minimum length of chromosomes is observed during which one of the following phases?
(a) Anaphase (b)Metaphase
(c) Prophase (d) Telophase

20. Lamp brush chromosomes were first observed by which one of the following scientist?
(a) Darlington (b) Flemming.
(c) Strausberger (d) Waldeyer

21. The gametic chromosome number of a true diploid species is termed as which one of the following?
(a) Basic number (b) Haploid number
(c) Somatic number (d) None of these

22. The term genetic RNA refers to which one of the following?
(a) The RNA present in mitochondria.
(b) The RNA that carries genetic message.
(c) The RNA that helps gene regulation in lac-operon.
(d) None of these

23. Shape of chromosome is determined by the position of which one of the following phase?
(a) Centromere (b) Chromomere
(c) Telomere (d) None of these

24. In a haploid nucleus, a single set of chromosomes carries total genetic information. Which one of the following termed is?
(a) Chromosome (b) Gene
(c) Genome (d) Genomics

25. Primary constriction is also called as which one of the following?
(a) Centromere (b) Chromomere
(c) Kinetochore (d) Both (b) and (c)

26. Which of the following is not a part of the chromosome?
(a) Centromere (b) Chromomere
(c) Spindle fiber (d) Telomere

27. If chromosome is ____________ the position of a centromere in holokinetic.
(a) Diffused (b) Median
(c) Sub-terminal (d) Terminal

28. Special chromosome refers to which one of the following chromosome?
(a) Accessory chromosome (b) Lampbrush chromosome
(c) Polytene chromosome (d) All of the above

29. Normal chromosome refers to which one of the following chromosome?
(a) A-chromosome (b) B-chromosome
(c) Iso chromosome (d) Ring chromosome

30. Genophore refers to which one of the following?
(a) DNA and protein in prokaryotes.
(b) DNA and RNA in prokaryotes.
(c) DNA in prokaryotes. (d) RNA in prokaryotes.

31. A chromosome with two identical arms is termed which one of the following?
(a) A-chromosome (b) B-chromosome
(c) Accessory chromosome (d) Iso chromosome

32. B-chromosome is known as which one of the following, except?
(a) Accessory chromosome (b) Lampbrush chromosome
(c) Polytene chromosome (d) Supernumerary chromosome

33. Which one of the following is a unit of a parent and inherited as a complete set (n) of chromosomes?
(a) Chromosome (b) Gene
(c) Genome (d) Genomics

34. Three genomes (A, B, (D) comprise the polyploid series of wheat. which one of the following is debatable?
(a) A genome (b) B genome (c) D genome (d) None of these

35. A chromosome with diffused centromere is termed as which one of the following?
(a) Acrocentric (b) Holokinetic
(c) Telocentric (d) None of these

36. Which one of the following term is used for a chromosome with a median position of centromere?
(a) A crocentric (b) Meta centric
(c) Sub-matacentric (d) Telocentric

37. A metacentric chromosome will assume: which one of the following at anaphase?
(a) J Shape (b) L Shape
(c) V Shape (d) Rod Shape

38. The chromosome with sub-terminal position of centromere is termed as which one of the following?
(a) Acrocentic (b) Holokinetic
(c) Telocentric (d) None of these

39. Centromere is also known as which one of the following?
(a) Chromomere (b) Primary constriction
(c) Secondary constriction (d) All of the above

40. In the same chromosome or genome, the occurrence of a segment more than once is termed as which one of the following terms?
(a) Deletion (b) Duplication
(c) Inversion (d) None of these

41. The types of ribosomes found in prokaryotic cells are which one of the following
(a) 60 S (b) 70 S (c) 80 S (d) 100 S

42. An organism with four genomes derived from hybridization of different species. Usually, in forms that become established, two of the four genomes are from one species and two are from another species is?
(a) Allotetraploid (b) Autotetraploid
(c) Di haploid (d) None of these

43. At anaphase, movement of chromosome is associated with which one of the following?
(a) Centromere (b) Chromomere
(c) Telomere (d) None of these

44. Chromosome Puffs are sites of which one of the following synthesis?
(a) Protein synthesis (b) DNA synthesis
(c) RNA synthesis (d) None of these

45. In each cell of an organism, which one of the following refers to the specific allelic composition of a particular gene or set of genes and may also refer to the entire genome?
(a) Genotype (b) Karyotype
(c) Phenotype (d) None of these

46. In which one of the following chromosomes, puffs are found?
(a) B- chromosome (b) Iso-chromosome
(c) Lampbrush chromosome (d) Polytene chromosomes

47. Which one of the following is a pair of synapsed or associated homologous chromosomes that may or may not have undergone the duplication process to form a group of four chromatids?
(a) Bivalent (b) Quadrinatant
(c) Tetrad (d) Univalent

48. Which one of the following is termed for a polyploidy having multiple and identical or nearly identical sets of chromosomes (genomes)?
(a) Allopolyploid (b) Autopolyploid
(c) Segmental polyploid (d) None of these

49. An exact multiple of the monoploid (*n*) with one genome in an organism or cell which has chromosome number is known as which one among the following?
(a) Aneuploid (b) Dihaploid
(c) Euploid (d) Both (b) and (c)

50. The term allele was coined in 1909 by which one of the following synthesis?
(a) Bridges (b) Johannsen
(c) Mendel (d) Morgan

51. Lampbrush chromosome is a particular type of chromosome shape, usually considered with which one of the following?
(a) Chromosome regions may also occur in plants e.g. Phaseolus beans
(b) Found in the diplotene stage of animals, also known as puffed chromosome.
(c) The loops of DNA strands form a lampbrush-like shape.
(d) All of the above

52. Which one of the following are true about chromosomes?
(a) A cell or nucleus having two sets of homologous chromosomes is described by diploid.
(b) In a polyploid series, monoploid is term for basic chromosome number.
(c) In the nucleus gene, tetraploid is term for four sets of chromosomes.
(d) All of the above

53. Which one of the following is normal pattern by which deficiency, duplication, inversion, translocation, aneuploid, polyploidy, or any change occurs in abnormal structure or number of chromosomes?
(a) Chromosome aberration (b) Euploidy
(c) Dublication (d) None of these

54. Loops are found in which one of the following chromosomes?
(a) Iso chromosomes
(b) Lampbrush chromosomes
(c) Polytene chromosomes (d) None of these

55. Which one of the following chromosomes is large diplotene chromosomes which are active sites of transcription, present in oocyte nuclei, particularly in amphibians and have extended regions called loops?
(a) Iso chromosome (b) Lampbrush chromosome
(c) Polytene Chromosome (d) None of these

56. An allele is a number of alternative forms of the same gene or same genetic locus (generally a group of genes). The allele is also known as which one of the following?
(a) Factor (b) Hereditary determiner
(c) Both (a) and (b) (d) None of these

57. Which one of the following has one of the largest genomes with over 150,000 genes as compared to rice with about 40,000 genes?
(a) Common wheat (b) Maize
(c) Pulses (d) None of these

58. A chromosome that contributes to the determination of sex is which one of the following?
(a) Heterogenic (b) Heteromeric
(c) Heteroplasmic (d) Idiochromosome

59. Which one of the following is a scientific study of genes and their role in the structure, growth, health, and disease of plant?
(a) Genomics (b) Genome
(c) Both (a) and (b) (d) None of these

60. Which one of the following cell organelles is the location of the cell's genome?
(a) Cytoplasm (b) Golgi apparatus
(c) Mitochondria (d) Nucleus

61. Retroviruses convert which one of the following with enzyme reverse transcriptase, then back into mRNA by normal transcription?
(a) DNA into RNA genome (b) RNA genome into DNA
(c) Both (a) and (b) (d) None of these

62. The most recent achievement is the successful completion of which one of the following projects?
(a) Human genome project (HGP)
(b) Rice genome project (RGP)
(c) Wheat genome project (WGP)
(d) Plant genome project (PGP)

63. Genome annotation is the process of attaching biological information to sequences. Which of the following steps is/are the main steps constitute of genome annotation?
(a) Attaching biological information to these elements.
(b) Identifying elements on the genome, a process called gene prediction.
(c) Identifying portions of the genome that do not code for proteins.
(d) All of the above

64. The genomes A and D come from which one of the following polyploid series of wheat?
(a) Triticum aestivum and Triticum monococcum respectively.
(b) Triticum monococcum and Triticum tauschii respectively.
(c) Triticum tauschii and Triticum aestivum respectively.
(d) Triticum tauschii and Triticum monococcum respectively.

ANSWERSHEET

10. Genome and Genomic Organizations

1. d	2. a	3. d	4. c	5. d.	6. b	7. a
8. a	9. c	10. c	11. b	12. a	13. c	14. b
15. d	16. c	17. b	18. a	19. a	20. b	21. a
22. b	23. a	24. c	25. a	26. c	27. a	28. d
29. a	30. b	31. d	32. b	33. c	34. b	35. b
36. b	37. c	38. a	39. b	40. b	41. b	42. a
43. a	44. c	45. a	46. d	47. a	48. b	49. d
50. b	51. d	52. d	53. a	54. b	55. b	56. c
57. a	58. d	59. a	60. d	61. b	62. a	63. d
64. b						

11

Stress and Quality Breeding

1. The main aims of quality breeding in plant are to produce which one of the following?
(a) Increased quality and yield of the crop.
(b) Increased tolerance of herbicides.
(c) Both (a) and (b) (d) None of these

2. Quality traits that breeders have tried to incorporate into crop plants for resistance to viruses, fungi and bacteria. Which one of the following is also tried by breeders quality breeding?
(a) Increased tolerance to insect pests.
(b) Increased tolerance of environmental pressures.
(c) Both (a) and (b) (d) None of these

3. CAM mechanism is observed in plants having which one of the following stress?
(a) Oxidative stress (b) Salt stress
(c) Water stress (d) None of these

4. Cereals tend to be low in which of the following amino acids?
(a) Lysine (b) Threonine
(c) Tryptophan (d) Both (a) and (b)

5. Development and use of which one among the following resistant cultivars is beneficial to crop production in areas of marginal or erratic rainfall regimes?
(a) Water resistant (b) Stress resistant
(c) Drought resistant (d) Salt resistant

6. Breeders need to develop new plant types that can resist various biotic stresses is
(a) Diseases (b) Insect pests
(c) Water (d) Both (a) and (b)

7. Forage breeders are very interested in improving feed quality by which one of the following for livestock?
   (a) High yielding and High digestibility
   (b) High nutritional profile and High yielding
   (c) High digestibility and High nutritional profile
   (d) None of these

8. Legumes tend to be low in which of the following amino acids?
   (a) Cysteine (b) Lysine
   (c) Methionine (d) Both (a) and (c)

9. Rice is a major world food, lacks of which of the following vitamins?
   (a) Pro-vitamin A (b) Vitamin B Complex
   (c) Vitamin C (d) None of these

10. Which one of the following is mainly responsible for greenhouse effect?
    (a) CO (b) CO_2 (c) O_2 (d) SO_2

11. A new plant types are developed by breeders that can resist salt, drought, heat, cold stresses. These are the examples of which one of the following type of stresses?
    (a) Abiotic stresses (b) Biotic stresses
    (c) Environmental stress (d) None of these

12. To overcome oxidative stress, which of the following enzyme is used?
    (a) Glucose-6-phosphate dehydrogenase (G6PD or G6PDH)
    (b) Lactate dehydrogenase (LDH or L(D)
    (c) Plesiochronous digital hierarchy (PDH)
    (d) Superoxide dismutases (SO(D)

13. Roundup Ready soybean introduced in which one of the following years is?
    (a) 1994 (b) 1995 (c) 1996 (d) 1997

14. In which one of the following years, roundup Ready wheat was developed?
    (a) 2002 (b) 2003 (c) 2004(d) 2005

15. Pulses are important food crops due to their high content of which one of the following?
    (a) Essential amino acid (b) Minerals
    (c) Protein (d) Both (a) and (c)

16. FlavrSavr tomato developed in which one of the following years as first genetically modified food produced for the market.
    (a) 1991 (b) 1994 (c) 1996(d) 2000

17. Jones developed first commercial hybrid maize in which one of the following years?
(a) 1915 (b) 1916 (c) 1917 (d) 1918

18. Which one of the following is a driving force for the process of 'passive absorption of water in roots'?
(a) Activity of aquaporins (b) Root pressure
(c) The increase in imbibitional pressure in root cells.
(d) Transpiration in leaves

19. The use of mutation breeding and biotechnology is to improve which one of the following stresses?
(a) Tolerance to drought and salinity
(b) Water use efficiency
(c) Both (a) and (b) (d) None of these

20. The environmental stresses resulting from drought, temperature, salinity etc. are the major limiting factors in crop production. Which one among the following is a main abiotic stress that limits crop production?
(a) Drought stress (b) Salinity stress
(c) Temperature stress (d) None of these

21. The absence of adequate moisture necessary for a plant to grow normally and complete its life cycle is known as
(a) Drought stress (b) Salinity stress
(c) Temperature stress (d) None of these

22. Drought stress affects crop growth and yield during which of the following developmental stages.
(a) Fertilization and Embryogenesis
(b) Reproductive organs and Gametogenesis
(c) Seed development (d) All of the above

23. Which one of the following is considered with effects of drought stress on pollination traits including decrease number of ovules fertilized and developed?
(a) Decrease megagametophyte Fertility
(b) Decrease number of pollen grain availability
(c) Increase number of pollen grain sterility
(d) All of the above

24. Increase pollen sterility, ovary abortion and inhibits cell division in the embryo are the role of which one of the following biochemical contents in response to drought stress?
(a) Abscisic acid (AB(A) (b) Gibbereline (G(A)
(c) Cytokininss (CKs) (d) None of these

25. Which one of the following is/are true statements about drought stress?
(a) Development of the ovary is one of the most vulnerable phases in response to drought stress.
(b) Flower attractiveness to pollinators can be negatively affected by drought stress.
(c) Flowering is one of the most important growth stage affected by drought stress.
(d) All of the above

26. The reduction in number of spikes per plant under which one of the following stress due to the increase in the number of sterile spikes per plant and the decrease in the number of fertile spikes per plant in six-row barley?
(a) Cold Stress (b) Drought stress
(c) Heat stress (d) Salinity stress

27. Stresses trigger to which one of the following plant responses?
(a) Altered gene expression (b) Cellular metabolism
(c) Changes in growth rates and crop yields
(d) All of the above

28. Plant Response to which one of the following stress arises from an excess or deficit in the physical or chemical environment?
(a) Abiotic Stress (b) Biotic Stress
(c) Both (a) and (b) (d) None of these

29. When the oxygen supply to the tissue is inadequate, the condition is known as which one of the following?
(a) Apnea (b) Asphyxia
(c) Dyspnea (d) Hypoxia

30. Resistance or sensitivity of plants to stress depends on which one of the following?
(a) Species (b) Genotype
(c) Development age (d) All of the above

31. Which one of the following is/are the response to changes in gene expression to stress?
(a) A stress response is initiated when plants recognizes stress at the cellular level.
(b) Changes in gene expression may modify growth and development and even influence reproductive capabilities.
(c) Stress recognition activates signal transduction pathways that transmit information within the individual cell and throughout the plant.
(d) All of the above

32. Regulation of plant stress is responses by which one of the following, except?
(a) Abscisic acid (AB(A) (b) Auxins
(c) Ethylene (d) Jasmonic acid

33. Environmental conditions can lead to water deficit, except?
(a) Drought (b) High temperatures
(c) Hypersaline conditions
(d) Transient loss of turgor at midday

34. A biochemical mechanism that helps plants acclimate to dry and saline conditions is known as
(a) Drought Tolerance (b) Osmotic adjustment
(c) Salinity Tolerance (d) Both (a) and (c)

35. Stress results from conditions promoting the formation of active oxygen species that damage or kill cells is known as
(a) Heat Stress (b) Oxidative Stress
(c) Reactive oxygen species (ROS)
(d) None of these

36. Qualitative variation like dwarf growth, monogermous fruits and Quantitative variation like yield/ha, harvest index, nutritional efficiency, rate of photosynthesis are the breeding aims which one of the following?
(a) Quality traits (b) Resistance traits
(c) Yield traits (d) None of these

37. The typical response to which one of the following stress is decrease in the synthesis of normal proteins, accompanied by an accelerated transcription and translation of new proteins?
(a) Heat stress (b) Oxidative stress
(c) Drought stress (d) None of these

38. Which one of the following environmental conditions that can cause stress, except?
(a) Adequate mineral in the soil (b) Excessive soil salinity
(c) High or low temperatures (d) Water-logging and drought

39. There are many reasons by which pathogen failure to cause disease, except?
(a) Defence mechanisms are activated such that the invasion remains localized.
(b) Environmental conditions change and the pathogen perish.
(c) The plant does not possess to preformed structural barriers or toxic compounds.
(d) The plant species attacked is unable to support the life-strategy of the pathogen.

40. **Which one of the following is the negative effects of ozone on plants, except?**
(a) Accelerated senescence
(b) Increased rates of photosynthesis
(c) Leaf injury and reduced crop yield
(d) Reduced growth of shoots and roots

41. **Pathogen attack strategies in which the plant cells are killed are known as which one of the following?**
(a) Biotrophy (b) Hemibiotrophy
(c) Necrotrophy (d) None of these

42. **Which one of the following term is used for the plant cells remain alive after pathogen attack?**
(a) Biotrophy (b) Hemibiotrophy
(c) Necrotrophy (d) None of these

43. **The pathogen initially keeps cells alive but kills them at later stages of infection is called as?**
(a) Biotrophy (b) Hemibiotrophy
(c) Necrotrophy (d) None of these

44. **An example of 'Vitamin A' rich transgenic plant is which one of the following?**
(a) Bt cotton (b) Flay saur tomato
(c) Golden rice (d) Vaccinated potato

45. **Tryptophan and Lysine are deficient amino acid in which one of the following crop?**
(a) Corn (b) Rye
(c) Soybean (d) Both (a) and (b)

46. **Which of the following is the tissue that is most important in plant survival during droughts and why?**
(a) Epidermal tissue, because it is covered with a waxy cuticle and includes drought responsive stomata
(b) Ground tissue, because water can be stored in the sclerenchyma for use during droughts
(c) Vascular tissue, because the phloem keeps the plant supplied with water
(d) Vascular tissue, because the xylem can store considerable amounts of water

47. Which one of the following compounds, may cause adverse or toxic effects and alter normal pattern of growth and development in ecosystems?
(a) Non-methane hydrocarbons (NMHCs)
(b) Peroxy acetyl nitrate (PAN)
(c) Volatile organic compounds (VOCs)
(d) All of the above

48. Changes in crop quality due to O3 exposure have been studied in a limited number of crops. Which one of the following is true regarding this?
(a) Decrease reducing sugars and increase Vit.-C content in potato tubers.
(b) Reduce the oil, protein, and carbohydrate contents in rape seeds.
(c) Reduced yield but increased grain protein concentration in wheat crops.
(d) All of the above

49. Salt tolerant crops are developed by use of which one of the following gene?
(a) Heat Shock Proteins (HSP) gene
(b) Late Embryogenesis Abundant (LE(A) gene
(c) Both (a) and (b) (d) None of these

50. Which one of the following is true regarding stress, except?
(a) Abiotic stress may not change in gene expression and cellular metabolism.
(b) Cotton is highly sensitive to environmental stresses.
(c) Leaf area development is highly sensitive to temperature.
(d) Soil salinity is abiotic stresses reducing agricultural productivity.

51. Qualitative variation like linolenic acid, large seeds and Quantitative variation like protein content, cooking ability, and digestibility are the breeding aims which one of the following?
(a) Quality traits (b) Resistance traits
(c) Yield traits (d) None of these

52. Which one of the following term is used for the ratio between the yields per unit of resource water or nutrients available to the plant?
(a) Acquisition efficiency (AE) (b) Use efficiency (UE)
(c) Utilization efficiency (UtE) (d) None of these

53. An ability of a given genotype to absorb certain nutrients or soil water is known as which one of the following?
(a) Acquisition efficiency (AE) (b) Use efficiency (UE)
(c) Utilization efficiency (UtE) (d) None of these

54. Which one of the following is refers as an ability of the genotype to produce biomass or grain using the resource that has been absorbed?
(a) Acquisition efficiency (AE) (b) Use efficiency (UE)
(c) Utilization efficiency (UtE) (d) None of these

55. Except, which one of the following has/have been recognized that plants use to cope with stress?
(a) Genotype has moderate suitability in most environments.
(b) Genotype is adapted to the specific environment.
(c) Never stimulate the production of alternative phenotypes.
(d) Signals from the environment interact with the genotype.

56. In vegetative tissues which one of the following is synthesized de novo mainly in response to drought and high salinity stresses?
(a) Abscisic acid (AB(A) (b) Gibbereline (G(A)
(c) Cytokininss (CKs) (d) None of these

57. Cavitation occurs in xylem of plants that are under moderate to severe which one of the following stress?
(a) Heat stress (b) Oxidative stress
(c) Water-deficit stress (d) None of these

58. A method used to screen for post-anthesis stress tolerance as destruction of the plant¡¯s photosynthetic system is called as chemical desiccation. Which one of the following most commonly chemical desiccants are used for cereals?
(a) Magnesium chlorate (b) Potassium iodide
(c) Sodium chlorate (d) All of the above

59. In terms of resistivity or sensitivity to which one of the following stress-related trait is response by the plant?
(a) Abiotic stress (b) Biotic stress
(c) Both (a) and (b) (d) None of these

60. Which one of the following tests is for seed determine the potential ability to develop into normal, healthy plants under a wide range of field conditions?
(a) Aging test (b) Cold test
(c) Paper-piercing test (d) Vigor test

61. In the presence of various soil microorganisms, which one of the following is a stress test that shows the performance of seeds in cool, moist soil?
(a) Aging test (b) Cold test
(c) Paper-piercing test (d) Vigor test

62. Which one of the following is a method to originally evaluate seed storability?
(a) Aging test (b) Cold test
(c) Paper-piercing test (d) Vigor test

63. Which one of the following is a stress test for seedling vigor utilizing sand covered by filter paper, through which the seedlings must emerge to be considered vigorous?
(a) Aging test (b) Cold test
(c) Paper-piercing test (d) Vigor test

64. Which one of the following test is conducted by planting the seeds in moist, unsterilized field soil, exposing them to cool (+5 to -10°(C) temperatures for about one week, then allowing them to germinate in the same soil at warmer conditions?
(a) Aging test (b) Cold test
(c) Paper-piercing test (d)Vigor test

65. Which one of the following stress tests is used to provide a more sensitive index of seed quality?
(a) Aging test (b) Cold test
(c) Paper-piercing test (d) Vigor test

66. Which one of the following test subjects unimbibed seeds to conditions of high temperature (+41°(C) and relative humidity (~100%) for short periods (3–4 days) the seeds are then removed from the stress conditions and placed under optimum germination conditions?
(a) Aging test (b) Cold test
(c) Paper-piercing test (d) Vigor test

67. Poor emergence, dormancy, seed deterioration, and other quality factors can be identified through which one of the following test?
(a) Aging test (b) Cold test
(c) Paper-piercing test (d) Vigor test

68. Stress refers as which one of the following
(a) A specific response by the plant to a stimulus that disturbs
(b) Interferes with the normal physiological equilibrium
(c) Both (a) and (b) (d) None of these

69. Which one of the following oligosaccharides is believed to play an important role in the resistance of plants to environmental stress by protecting membrane-bound proteins?
(a) Gentianose (b) Raffinose
(c) Rhamninose (d) Robinose

70. Mega-environment (ME) is defined as similar biotic and abiotic stresses, cropping system requirements, consumer preferences, and by a volume of production. The concept of Mega-environment (ME) was introduced by CIMMYT organization in which one of the following year to address the needs of diverse wheat growing areas of the world?
(a) 1986 (b) 1988 (c) 1989 (d) None of these

71. In stomata closure of water stressed plants, which one of the following is involved?
(a) Abscisic acid (ABA) (b) Cytokinin
(c) Gibberlin (d) None of these

72. Qualitative variation like monogenic resistance, tolerance to aluminum and Quantitative variation like polygenic resistance, drought tolerance are the breeding aims which one of the following?
(a) Quality traits (b) Resistance traits
(c) Yield traits (d) None of these

73. Pigeon pea and Common bean are deficient in which one of the following amino acids?
(a) Lysine (b) Methionine
(c) Tryptophan (d) None of these

74. Amino acid deficient in wheat and millet crops is?
(a) Cystine (b) Lysine
(c) Methionine (d) Tryptophan

75. Heat shock may arise in all the following conditions, except?
(a) Organs with reduced capacity for transpiration e.g. fruits
(b) Soil is warmed by the sun
(c) Stomata are partially or fully closed
(d) Transpiration is sufficient

76. Match the following stress resistance mechanisms with suitable match?

1. Avoidance	A. Alter their physiology in response stress.
2. Tolerance	B. Prevents exposure to stress.
3. Acclimation	C. Permit the plant to withstand stress.

(a) 1B 2C 3A (c) 1B 2A 3C
(b) 1C 2A 3B (d) 1C 2B 3A

77. Match the following Toxic substances occur in major food crops.

1. Alkaloids	A. Pulses
2. Cynogenic glucosides	B. Potatoes.
3. Trypsin inhibitors	C. Cassava
4. Steroidal alkaloids	D. Yam

(a) 1 C 2 A 3 D 4 B (c) 1 D 2 C 3 A 4 B
(b) 1.B 2.C 3.A 4 D (d) 1 A 2 D 3 C 4 B

78. Match the following essential amino acids that are deficient in selected major world food crops.

1. Rice (polishe(d)	A. Lysine, Methionine, Cystine, Threonine
2. Soybean	B. Methionine, Cystine
3. Peanut	C. Lysine, Threonine
4. Potato	D. Methionine, Cystine, Valine

(a) 1B 2C 3D 4A (c)1C 2D 3A 4B
(b) 1C 2D 3B 4A (d) 1D 2C 3A 4B

79. Match the following crops with their suitable test.

1. Soybean	A. Cool Germination Test
2. Maize	B. Alakoid test
3. Lupins	C. Seed coat peroxidase test
4. Soybean	D. Cold Test
5. Cotton	E. Accelerating Ageing Test

(a) 1C 2D 3B 4E 5A (c) 1C 2B 3D 4E 5A
(b) 1C 2D 3A 4E 5B (d) 1C 2A 3D 4E 5B

80. There are various classes of heat shock proteins (HSPs). Match the following classes of HSPs with their correct location.

1. HSP100	A. Cytoplasm, Chloroplast, ER, Mitochondria
2. HSP90	B. Chloroplasts, Mitochondria
3. HSP70	C. Cytoplasm
4. HSP60	D. Cytoplasm, ER
5. smHSP	E. ER, Cytoplasm, Mitochondria

(a) 1B 2C 3D 4E 5A (c) 1C 2D 3E 4B 5A
(b) 1C 2D 3E 4A 5B (d) 1D 2E 3A 4B 5C

81. Match the following chemical composition of soybean with their appropriate composition.

1. Protein	A. 35%
2. Carbohydrate	B. 5%
3. Ash	C. 20%
4. Oil	D. 40%

(a) 1B 2C 3D 4A (c) 1C 2D 3B 4A
(b) 1C 2D 3A 4B (d) 1D 2A 3B 4C

82. Different crop plants show differential response to salinity. Match the following salinity with suitable crop plants response.

1. Highly tolerant crops	A. Rice, corn, foxtail millet, cow pea, peanut, sugarcane, tomato, potato, sweet potato, radish, alfalfa, cabbage
2. Moderately Tolerant crops	B. Citrus, straw berry, melon, peas, other legumes, apple, rajmabean, carrot, okra, onion (orange)
3. Moderately sensitive	C. Sugarbeat, sunflower, barley (grain), cotton, datepalm, asparagus
4. Extremely sensitive	D. Barley (Forage), rye, soghum, wheat, safflower, soy bean

(a) 1B 2C 3D 4A (c) 1C 2D 3B 4A
(b) 1C 2D 3A 4B (d) 1D 2A 3B 4C

ANSWERSHEET

11. Stress and Quality Breeding

1. c	2. c	3. b	4. c	5. d	6. d	7. c
8. d	9. a	10. b	11. a	12. d	13. c	14. c
15. d	16. b	17. c	18. d	19. c	20. a	21. a
22. d	23. d	24. a	25. d	26. b	27. d	28. a
29. d	30. d	31. a	32. b	33. b	34. b	35. b
36. c	37. a	38. a	39. c	40. b	41. c	42. a
43. b	44. c	45. d	46. a	47. d	48. d	49. c
50. a	51. a	52. b	53. a	54. c	55. c	56. a
57. c	58. d	59. a	60. d	61. b	62. a	63. c
64. b	65. d	66. a	67. a	68. c	69. b	70. b
71. a	72. b	73. c	74. d	75. d	76. b	77. c
78. c	79. a	80. c	81. d	82. b		

12

Seed Science and Technology

12.1 Seed Industry Development

1. There are ____ number of sections in the Indian Seed Act (1966).
(a) 16 (b) 20 (c) 25 (d) 27

2. Chairman of the CVRC, who is among the following?
(a) DDG (CS), ICAR
(b) Deputy. Secretary (seeds)
(c) Director of Agriculture of the concerned state
(d) Director of Research of the SAU

3. International Seed Testing Association (IST(A) was born during which one among the following years?
(a) 1869 (b) 1924 (c) 1960 (d) 1973

4. According to Feistritzer (1975), which one of the following is true about the roles of improved seed?
(a) A carrier of new technologies and basic tool for secured food supply.
(b) A medium for rapid rehabilitation of agriculture in cases of natural disaster.
(c) The principal means to secure crop yields in less favourable areas of production.
(d) All of the above

5. The CSTL is presently located at:
(a) Hyderabad (b) New Delhi
(c) Pantnagar (d) Varanasi

6. Chairman of SVRC, who is among the following?
(a) DDG (CS), ICAR
(b) Deputy. Secretary (seeds)
(c) Director of Agriculture of the concerned state
(d) Director of Research of the SAU

7. Which one of the following is necessary to obtain the expected dividends from the use of seeds of improved varieties?
(a) Assured high quality of seeds (b) Rapid multiplication
(c) Reasonable price (d) Timely supply

8. The Seed Act was passed and implemented in our country respectively, in the years of:
(a) 1960, 1963 (b) 1964 , 1968
(c) 1966 , 1969 (d) 1968, 1972

9. Which one of the following activity is not included under seed marketing?
(a) Post-sales service (b) Seed demand forecast
(c) Seed pricing (d) Seed quality control

10. ______________ of a crop variety is mandatory for inclusion in the seed chain.
(a) Certification (b) Notification
(c) Release (d) None of these

11. Which one of the following is a non-formal seed distribution system?
(a) Distribution by cooperatives
(b) Distribution by Government organization
(c) Distribution by Non-Govt. organization
(d) Farmer to farmer distribution

12. The major goal of seed technology is to increase agricultural production through the spread of good quality seeds of high yielding varieties. It aims at which one of the following?
(a) Rapid multiplication (b) Reasonable price
(c) Timely supply (d) All of the above

13. The Journal of *Seed Research* is published by which one of the following organizations?
(a) IARI (b) ISST (c) ISTA (d) NSC

14. Which one of the following committee recommended withdrawal of the state Governments from seed procurement and distribution system?
(a) Grow More Food Enquiry Committee (1952)
(b) Programme Evaluation Organization (1960)
(c) Seed Multiplication Review Team (1961)
(d) Seed Review Team (1968)

15. Which of the following is not a seed quality control organisation?
(a) CSTL (b) SCA (c) SFCI (d) SSTL

16. Royal Commission on Agriculture was constituted in which one of the following years?
(a) 1923 (b) 1925 (c) 1928 (d) 1929

17. The Seed Certification Agencies and Central Seed Certification Board were formed in the year:
(a) 1968 (b) 1970 (c) 1972 (d) 1976

18. Which of the following is the largest seed producing organization?
(a) CRRI (b) IARI (c) NSC (d) SFCI

19. Which one of the following agency deal with all problems related to seed certification and to co-ordinate the work of sate seed certification agency?
(a) Central Seed Certification Board (CSC(B)
(b) Central Seed Testing Laboratory (CSTL)
(c) State Seed Certification Agency (SSC(A)
(d) State Seed Testing Laboratory (SSTL)

20. Which one of the following system seems to be more efficient in sales promotion activity for seed?
(a) Field demonstration (b) Hoardings
(c) New paper (d) Radio

21. The private sectors mainly deal with production of which one of the following seeds?
(a) Low value-high volume seeds
(b) High value –low volume seeds
(c) Vegetable seeds
(d) All type of seeds

22. Royal Commission on Agriculture was constituted and it made on which of the following recommendations for introduction and spread of improved varieties?
(a) Private seed growers should be given encouragement.
(b) Separate organization should be there within agriculture to deal with seed testing and distribution.
(c) The seed distribution should be made through co-operative societies, dept. of agriculture and seed growers.
(d) All of the above

23. The National Seed Project (NSP-I) was implemented in India in the year of:
(a) 1975-76 (b) 1977-78
(c) 1980-81 (d) 1981-82

24. The ISST stands for which one of the following?:
(a) International Society of Seed Technology
(b) Indian Society of Seed Technology
(c) International Society of Seed Trade
(d) Indian Seed Survey Team

25. In 1944, which of the following committee was/were constituted?
(a) Dr. Burns Committee
(b) Famine Enquiry Committee
(c) Food Grain Policy Committee (d) All of the above

26. In producing seed, additional expenditure on which one of the following activity is to be made in comparison to commercial crop production.
(a) Cost of fertilizers (b) Land preparation
(c) Roguing (d) Weeding and plant protection

27. The AOSA was established in the year:
(a) 1906 (b) 1908 (c) 1909 (d) 1910

28. Which of the following activity is not included under seed technology in narrow sense?
(a) Seed certification (b) Seed marketing
(c) Seed production (d) Variety development

29. All India Seed Growers, Merchants and Nurserymen's Association was formed in which one of the following years with the objective of rapid development of vegetable Seed Industry?
(a) 1944 (b) 1945 (c) 1946 (d) 1947

30. The State Farms Corporation of India Limited (SFCI) was established in which one among the following years?
(a) 1963 (b) 1966 (c) 1968 (d) 1969

31. Put the right sequence of the following events:
(1) Seed multiplication 2) Variety release 3) Seed distribution 4) Notification
(a) 1-2-3-4 (b) 1-3-2-4 (c) 2-4-3-1 (d) 2-4-1-3

32. Grow More Food Enquiry Committee (GMFE(C) was constituted in which one of the following years that recommended seed multiplication and distribution schemes?
(a) 1951 (b) 1952 (c) 1955 (d) 1958

33. The journal *Seed Science & Technology* is published by which one of the following organizations?
(a) AOSA (b) ICAR (c) ISST (d) ISTA

34. There are ____ number of parts in The Seeds Rules (1968).
(a) 10 (b) 11 (c) 12 (d) 15

35. The journal *of Seed Technology* is published by which one of the following organizations?
(a) AOSA (b) IARI (c) ISST (d) ISTA

36. Generally which one of the following period is regarded as golden period for Agricultural Development and Research?
(a) First Five Year Plan (1951-56)
(b) Fourth Five -year plan (1969-74)
(c) Second five -year Plan (1956-61)
(d) Third Five Year Plan (1961-66)

37. Which of the following is not the ISTA International Seed Analysis Certificate?
(a) Blue (b) Green (c) Green (d) Orange

38. According to The PPV & FR Act (2001), a field crop variety remains protected for how many years?
(a) 8 (b) 10 (c) 15 (d) 18

39. Govt. of India was launched high yielding variety (HYV) program in which one of the following years?
(a) 1964 (b) 1965 (c) 1966 (d) 1967

40. The National Seed Project (NSP) has functioned in our country in how many phases?
(a) Two (b) Three (c) Four (d) None of these

41. Tarai Development Corporation (TD(C) was established in which one of the following years with the assistance from World Bank?
(a) 1952 (b) 1961 (c) 1966 (d) 1969

42. Which of the following is a public sector seed enterprise?
(a) Indo-American Seeds (b) Mahyco
(c) Monsanto (d) None of these

43. Which of the following benefit is not provided to the farmers under The PPV & FR Act (2001)?
(a) Right to use (b) Right to save
(c) Right to exchange
(d) Right to sell branded seed commercially

44. Indian Society of Seed Technology (ISST) was formed in which one of the following years that provide an opportunity for exchange of experiences and scientific knowledge to persons who engaged in seed industry?
(a) 1961 (b) 1966 (c) 1971 (d) None of these

45. Which of the following is not a seed producing organization?
(a) CRRI (b) CSTL (c) IARI (d) SAU

46. The Seeds (control) Order 1983 mostly regulates by which one of the following?
(a) Marketing of seed (b) Processing of seed
(c) Production of seed (d) Testing of seed

47. All India Coordinated Maize Improvement Project (AICMIP) was started in which one of the following years with the collaboration with Rockfeller foundation of USA with multidisciplinary approach?
(a) 1948 (b) 1967 (c) 1957 (d) 1966

48. Which of the following is different in respect of its activity?
(a) CRRI (b) DRR (c) ICRISAT (d) IRRI

49. The first seed testing laboratory in our country was established in which one of the following the year?
(a) 1960 (b) 1961 (c) 1963 (d) 1966

50. National Seeds Corporation (NS(C) was establishment in which one of the following years?
(a) 1962 (b) 1963 (c) 1964 (d) 1966

51. Which of the following is not a statutory function of NSC?
(a) Seed marketing (b) Seed processing
(c) Seed production (d) Variety development

52. The crop which contribute maximum to Indian Green Revolution is which one of the following?
(a) Maize (b) Pearl millet
(c) Rice (d) Wheat

53. Seeds (control) Order was passed in which one of the following five year plan that declares seeds are as an essential commodity?
(a) Third Five Year Plan (1961-66)
(b) Fourth Five -year plan (1969-74)
(c) Fifth five-year Plan (1974-77)
(d) Sixth five-year Plan (1980-85)

54. Which of the following is different in respect of its activity?
(a) AOSA (b) CSTL (c) ISTA (d) NSC

55. The Seed Inspectors are empowered to draw and dispatch seed samples for analysis under section ____ of The Seed Act, 1966.
(a) 13 (b) 14 (c) 15 (d) 16

56. Match the following releasing time with the varieties.

1. 1961 A. First sorghum hybrid CSH-1 from New Delhi
2. 1965 B. First four maize hybrids
3. 1964 C. First bajra hybrid HB -1 from Punjab

(a) 1C 2B 3A (c) 1B 2A 3C
(b) 1C 2A 3B (d) 1B 2C 3A

57. Match the following time with their suitable Committee when they are formed.

1. 1944 A. Sir John Russell Committee
2. 1940 B. ICAR Committee
3. 1937 C. Dr. Burns Committee

(a) 1C 2B 3A (c) 1B 2A 3C
(b) 1C 2A 3B (d) 1B 2C 3A

12.2 Seed Biology

1. For commercial cultivation produced with the involvement of homozygous white (ww) female and homozygous brown male (WW) in case of incomplete dominance, which one of the following colour of the hybrid seed of pigeon pea is provided?
(a) Brown (b) Brown with white streak
(c) Light brown (d) White

2. During the development of dicot seeds, which of the following mainly contributes to the formation of suspensor?
(a) Antipodal cells (b) Apical cell
(c) Basal cell (d) Egg cell

3. Soybean varieties may be classified as brown and black based on hilum colour. To work out the genetics, crosses were made between both the expressions. To determine the dominance the expression has to be observed on the seed in the generation of which one of the following?
(a) F_1 (b) F_2 (c) F_3 (d) None of these

4. The seed consists of which one of the following parts, except?
(a) Apical cell (b) Embryo
(c) Seed coat (d) Storage tissues

5. The inner coat of seed refers to which one of the following?
(a) Exine (b) Intine (c) Tegmen (d) None of these

6. The edible portion of litchi fruit is known as which one of the following?
(a) Aril (b) Endocarp (c) Endosperm (d) Mesocarp

7. All the haploid cells present in pollen tube, have same genetic constitution because the last cell division to form the male gamete cells is termed as which one of the following?
(a) Meiosis (b) Mitosis
(c) Reduction (d) Unknown

8. Which of the following seed absolutely does not possess any food reserve?
(a) Amaranthus seed (b) Grass seed
(c) Orchid seed (d) True Potato Seed

9. Source sink mechanism in seed was first given by which one of the following scientist?
(a) Mendal (b) Pulliam
(c) Ross (d) Whitteker

10. The apomictic seeds are usually known as which one of the following?
(a) Diploid (b) Haploid
(c) Triploid (d) Both (a) and (b)

11. In pollen tube, which one of the following is the number of haploid cells?
(a) 1 (b) 2 (c) 3 (d) 5

12. The process of double fertilization involves which of the following events?
(a) Fertilization and Apomixes
(b) Fertilization and triple fusion
(c) Pollination and Fertilization
(d) Pollination and Triple fusion

13. The path that has to be follow by pollen tube is directed by which one of the following?
(a) First egg cell (b) Generative cell
(c) Second egg cell (d) Tube nucleus

14. The whitish scar present on the lateral side of the seed representing attachment of the seed to placenta is
(a) Hilum (b) Micropyle
(c) Puckering (d) Raphae

15. In which of the following group of crops, true seeds are the agricultural seeds?
(a) Cereals (b) Fibre crops
(c) Oilseeds (d) Pulses

16. Which one of the following term is used for reaching of pollen upto stigma?
(a) Anthesis (b) Fertilization
(c) Germination (d) Pollination

17. Which one of the following is region of integumentary origin and attachment opposite to micropyle with prominent expression in chickpea?
(a) Chalaza end (b) Embryosac
(c) Hilum (d) Micropyle

18. A living part of a seed is, except which of the following?
(a) Aleurone layer (b) Cotyledon
(c) Embryo (d) Endosperm

19. The nuclear type of endosperm development occurs in about ____% angiospermic plants.
(a) 45 (b) 56 (c) 62 (d) 70

20. In the Gramineae family, the cotyledon of seed is
(a) Coleoptile (b) Embryo
(c) Endosperm (d) Scutellum

21. A living part of a seed among which of the following?
(a) Cotyledon (b) Endosperm
(c) Tegmen (d) Testa

22. Except which one of the following, seeds can be produced?
(a) Angiosperms (b) Ferns
(c) Flowering plants (d) Gymnosperms

23. The liquid endosperm of coconut is rich in which of the following growth promoter?
(a) Auxin (b) Cytokinins
(c) Ethylene (d) Gibberellins

24. Which one of the following is the leguminous crop with well-developed endosperm?
(a) Chickpea (b) Fenugreek
(c) Rice bean (d) Soybean

25. Endogenously controlled but environmentally imposed temporarily suspension of growth accompanied by reduced metabolic activity and relatively independent of ambient environmental conditions is known as which one of the following?
(a) Dead seed (b) Dormancy
(c) Non germinable (d) Quiescence

26. Which one of the following term is used for single seeded fruit with fused seed coat and fruit wall?
(a) Achene (b) Berry
(c) Caryopsis (d) Schizocarp

27. The stage of seed development when dry matter accumulation is maximum is called which one of the following?
(a) Field maturity (b) Harvestable maturity
(c) Physiological maturity (d) All of the above

28. Which one of the following is the scar on seed, shows the attachment of ovule with ovary?
(a) Carauncle (b) Hilum
(c) Micropyle (d) Raphae

29. Which one of the following term is used for single seeded fruit without fused seed coat and fruit wall used as seed?
(a) Achene (b) Berry
(c) Caryopsis (d) Schizocarp

30. Outgrowth of the hilum region which restricts water movement into and out of some seeds is termed as which one of the following?
(a) Carauncle (b) Coleorhiza
(c) Raphae (d) Strophiole

31. The moisture content of seeds at the physiological maturity stage ranges between:
(a) 20-30 (b) 30-40 (c) 40-50 (d) 50-60

32. An example of mitotic cell divisions is which of the following would most likely?
(a) Cross section of a leaf
(b) Cross section of an anther
(c) Cross section of muscle tissue
(d) Longitudinal section of a shoot tip

33. Which one of the following is sort out the odd crop among chickpea, mung bean, field pea and lentil based on germination?
(a) Chickpea (b) Field pea
(c) Lentil (d) Mung bean

34. Mungbean, sunflower, soybean and cotton have which one of the following similarity with each other?
(a) Epigeal germination (b) Oil seed crops
(c) Pulse crops (d) Self-pollinated

35. The process of seed development from the unfertilized egg cell is termed as which one of the following terms?
(a) Amphimixis (b) Apogamety
(c) Apospory (d) Parthenogenesis

36. What is the germination type of seed of all the designated oil seed crops?
(a) Epigeal (b) Hypo-epigeal
(c) Hypogeal (d) Viviaprous

37. Without fertilization, formation of true seed is termesd as which one of the following?
(a) Apomixis (b) Parthenogenesis
(c) Amphimixis (d) sexual reproduction

38. Which one of the following chromosome number is in cotyledon and endosperm, respectivly?
(a) 2n (b) 3n
(c) 2n and 3n (d) 3n and 2n

39. Polyembryonic seeds usually possess _____ number of fertilized embryo:
(a) One (b) Two (c) Three (d) Not fixed

40. Sorghum and Niger Seeds have one similarity because that both are refers as which one of the following?
(a) Dicot (b) Fruit (c) Monocot (d) Seed

41. The apical cell divides mitotically and forms a globular structure of diploid cells. How many numbers of diploid cells are?
(a) 5 (b) 8 (c) 16 (d) 10

42. After fertilization, formation of embryo from egg cell and endosperm is known as which one of the following?
(a) Cyto- differentiation (b) Genetic-differentiation
(c) Histo-differentiation (d) Myco-differentiation

43. In monocot seed, the cell layer separating embryo from the endosperm is called which one of the following?
(a) Aleurone (b) Epithelium
(c) Scutellum (d) Seed coat

44. In dicot seed, the part of embryo projecting out of cotyledon is called as which one of the following?
(a) Coleptile (b) Endosperm
(c) Plumule (d) Radicle

45. How many numbers of haploid cells are found in a *Polygonum* type embryosac?
(a) 2 (b) 3 (c) 6 (d) 8

46. Which one of the following option is correct for the genetic constitution and chromosome number of embryo and cotyledon?
(a) Different and Different (b) Different and Same
(c) Same and Different (d) Same and Same

47. The physiological maturity stage of seed is usually attained ho2w long days after fertilization?
(a) 15 (b) 20 (c) 30 (d) 45

48. A dicot seed germinates as epigeal due to fast growth in which one of the following?
(a) Cotyledon (b) Epicotyl
(c) Hypocotyl (d) Mesocotyl

49. Upto which one of the following stage, the development of monocot and dicot seed is same?
(a) Globular stage (b) Porembryo stage
(c) *Scutellar stage* (d) Torpedo stage

50. With the help of fertilization, embryo is formed between which one of the following?
(a) Egg cell and Sperm cell (b) Pollar nuclei and Egg cell
(c) Pollar nuclei and Sperm cell (d) None of these

51. Which of the following plant is propagated by root tuber?
(a) Potato (b) Radish
(c) Sweet potato (d) Yam

52. A viable seed not able to germinate because of unavailability of required environmental conditions is called as which one of the following?
(a) Dead seed (b) Dormant seed
(c) Hard seed (d) Quiescent

53. Archisporium haploid cells in embryosac, are arranged in the fashion of which one of the following?
(a) 1+2+1 (b) 2+2+2
(c) 3+2+3 (d) 3+3+3

54. Zygote is the growth of the cell formed by fusion of which pair of the following?
(a) Egg Cell and Sperm Cell (b) Egg Cell and Synergids
(c) Pollar Nuclei and Sperm Cell (d) Synergids and Sperm Cell

55. Tuberose is normally propagated by which one of the following materials?
(a) Bulbs (b) Corms (c) Rhizome (d) Sucker

56. Under favourable condition, a viable seed does not able to germinate is termed as which one of the following?
(a) Dead (b) Dormant
(c) Fresh ungerminated (d) Quiescent

57. At physiological maturity, which one of the following layer is formed between connections of ovule with ovary?
(a) Abscission layer (b) Glass layer
(c) Hard layer (d) Lime layer

58. Seed coat is made up of which one of the following?
(a) Diploid hybrid tissue (b) Diploid paternal tissue
(c) Diploid maternal tissue (d) Haploid maternal tissue

59. Gladiolus is normally propagated by which one of the following materials?
(a) Bulbs (b) Corms (c) Rhizome (d) Sucker

60. Which one of the following crop is with liquid endosperm?
(a) Buckwheat (b) *Cajanus cajan*
(c) Coconut (d) Rice

61. Outgrowth of funiculus, raphe, or integuments; or fleshy integuments or seed coat, a sarco-testa is termed as which one of the following except
(a) Aril (b) Elaiosomes
(c) Operculum (d) Strophiole

62. The plants in which flowers don't open are known as which one of the following?
(a) Chasmo-cleistogamous (b) Chasmogamous
(c) Cleistogamous (d) None of these

63. Which one of the following is dicot seed without cotyledon?
(a) Chickling vetch (b) *Cuscutta*
(c) Fenugreek (d) Lettuce

64. Hybrid and selfed seed of which one of the following crop can be distinguished at seed level with visual observation?
(a) Maize (b) Pigeonpea
(c) Rice (d) Sunflower

65. Hybrid seed produced by crossing between female of recessive yellow (gg) cotyledon colour with the male of dominant green (GG) cotyledon colour. The colour of the cotyledon of a hybrid seed will be
(a) Dark green (b) Green
(c) Light green (d) Yellow

66. Which of the following condition of flower enforces cross-pollination?
(a) Dichogamy (b) Dicliny
(c) Herkogamy (d) All of the above

67. Seed with endosperm and cotyledon both as store of food in which one of the following crops?
(a) Castor (b) Faba bean
(c) Fenugreek (d) Lettuce

68. Which one of the following is the contribution of female parent in endosperm with genetic constitution of YYy.
(a) YY (b) Yy (c) Yy (d) YYy

69. The ploidy level of endosperm is usually denoted by which one of the following?
(a) 2n (b) 3n (c) 4n (d) n

70. Which of the following statements is true?
(a) All agricultural seeds are true seeds.
(b) All flowering plants produce true seeds.
(c) All true seeds are agricultural seeds.
(d) Formation of true seed involves fertilization.

71. In which one of the following crops, the seed with endosperm, cotyledon and nucellus tissues as store of food?
(a) Beet (b) Castor
(c) Cucrbits (d) Tomato

72. All the seeds formed in a brinjal fruit, which one of the following are required for the number of pollen?
(a) Equal number of pollen (b) Half number of pollen
(c) One fourth number of pollen (d) One pollen

73. In which one of the following crop plants, normal root cuttings are used as agricultural seed?
(a) Pointed gourd (*Trichosanthes dioic*(a)
(b) Radish (*Raphanus sativus*)
(c) Sugarcane (*Saccharum officinarum*)
(d) Yam (*Dioscorea* alat(a)

74. True seed of sugarcane is known as which one of the following?
(a) Achene (b) Berry (c) Fuzz (d) Nut

75. True seed of potato is formed in which one of the following?
(a) Berry (b) Micro-tuber
(c) Pod (d) Tuber

76. The inbreeding depression is __________ in case of apomictic plants.
(a) High (b) Low (c) Very high (d) Zero

77. In a dormant seed, which one of the following is the ratio of growth promoter or growth inhibitor?
(a) 0 (b) <1 (c) >1 (d) 1

78. In which of the following crop seed, the seed coat and pericarp are fused?
(a) Cotton (b) Green gram
(c) Groundnut (d) Rice

79. Crop of ex-albuminous monocot seed with epigeal germination is which one of the following?
(a) Castor (b) Onion
(c) Potato (d) Sugarcane

80. During germination, the photo-synthetically functional cotyledon is seen in which one of the following crops?
(a) Chickpea (b) Field pea
(c) Lentil (d) castor

81. Which one of the following part of coconut is used as food?
(a) Cotyledon (b) Endocarp
(c) Endosperm (d) Mesocarp

82. Which of the following ecological factors have effect on seed development on crops?
(a) Light (b) Rainfall
(c) Temperature (d) All of the above

83. Ovule is attached with ovary by which one of the following?
(a) Chalaza (b) Funiculus
(c) Pollen tube (d) Raphae

84. Before harvesting, germination of mature seed on plant is called as which one of the following?
(a) Post-harvest sprouting (b) Quiescence
(c) Vivipary (d) Field emergence

85. Hypogeal germination occurs in which one of the following crops?
(a) Field pea (b) Mung bean
(c) Niger (d) Soybean

86. Epigeal germination occurs in which one of the following crops?
(a) Chickpea (b) Field pea
(c) Lentil (d) Mung bean

87. In cross-pollinated species, the pollen grains possess some special features like which of the following?
(a) Tiny in size (b) Sticky in nature
(c) Have wings or spine like outgrowth
(d) All of the above

88. Scar of chalaza end is prominent in the seed of which one of the following crops?
(a) Chickpea (b) Lentil
(c) Mung pea (d) Pigeonpea

89. Seeds of all the oilseed crops are germinated as which one of the following germination types?
(a) Epigeal (b) Hypogeal
(c) Epi-hypogeal (d) None of these

90. Coriander and Carrot are an example of which one of the following term that refers to the dried fruits, which are separated into two or more units at maturity?
(a) Achene (b) Endocarp
(c) Mezocarp (d) Schizocarp

91. For the germination and early growth of a dicotyledon seedling, the correct order of events is which one among the following?
(i) Cotyledon Digestion (ii) Imbibition
(iii) Root Growth (iv) Shoot Growth
(a) i–ii–iii–iv (b) ii–i–iii–iv
(c) iii–ii–i–iv (d) iv–ii–iii–i

92. In case of sporophytic incompatibility, the incompatibility reaction is determined by which of the following?
(a) Genotype of the female parent
(b) Genotype of the pollen
(c) Both (a) and (b) (d) None of these

93. Germination process of seed is initiated around a seed moisture content of:
(a) 20 % (b) 30 % (c) 40 % (d) 50 %

94. Moisture absorption by seeds from the soil is favoured by which of the following operation ?
(a) Hoeing (b) Laddering
(c) Ploughing (d) Weeding

95. Scarification of seed coat is done to break seed dormancy of which of the following categories?
(a) Anatomical (b) Morphological
(c) Physical (d) Physiological

96. To consider seed dormancy a desirable trait, which of the following statements is the best fit?
(a) Dormant seeds often possess better storability.
(b) Seed dormancy prevents field sprouting of seeds.
(c) Seeds can survive adverse climatic and soil conditions due to presence of dormancy.
(d) All of these

97. The seeds which germinate only in the absence of light are called:
(a) Negative photoblastic seeds (b) Non-photoblastic seeds
(c) Positive photoblastic seeds (d) None of these

98. Which one of the following scientist first evolved the viability nomograph in rice?
(a) Delouche (1973) (b) Harrington (1960)
(c) Roberts and Abdalla (1968) (d) Roberts and Roberts (1972)

99. Which one of the following scientist first put forth the viability equation?
(a) Delouche (1973) (b) Harrington (1960)
(c) Roberts and Abdalla (1968) (d) Roberts and Roberts (1972)

100. Which one of the following terms refer to the encapsulated somatic embryos?
(a) Artificial seed (b) Synseed
(c) Synthetic seed (d) All of the above

101. The concept of synthetic seed was first conceived by which of the following scientist?
(a) Gray et al. (1987) (b) Murasighe (1978)
(c) Redenbaugh et al. (1986) (d) Steward et al. (1958)

102. Which one of the following chemical is frequently used to induce desiccation tolerance of somatic embryos?
(a) Abscisic acid (b) High sucrose concentration
(c) Proline (d) Triazoles

103. A small, usually single-seeded, dry, indehiscent fruit formed from a single carpel termed as which one of the following?
(a) Achene (b) Endocarp
(c) Mezocarp (d) Schizocarp

104. Which one of the following term is used for inner layer of the fruit wall (pericarp); it may be hard and stony (peach pit), membranous (apple core), or fleshy (orange pulp)?
(a) Achene (b) Endocarp (c) Mezocarp (d) Schizocarp

105. In case of gametophytic incompatibility, the incompatibility reaction is determined by which of the following?
(a) Genotype of the female parent
(b) Genotype of the pollen
(c) Both (a) and (b) (d) None of these

106. Outgrowth of integument present on the seed of castor is known as which one of the following?
(a) Caruncle (b) Coleorhiza
(c) Endosperm (d) Strophiole

107. Which one of the following term is used for acquisition or acceleration of the ability to flower by chilling treatment?
(a) Articulation (b) Bolting
(c) Verbalization (d) Vernalization

108. From source to sink, sugar translocate in a seed of monocot found in which one of the following form?
(a) Raffinose (b) Stachyose
(c) Sucrose (d) Verbascose

109. The genetic male sterile lines are maintained by crossing them with which of the following?
(a) By selfing
(b) Dominant homozygous male fertile line
(c) Heterozygote male fertile line (d) None of these

110. In ex-albuminous seed, the food material is absorbed in the cotyledon with the help of which one of the following?
(a) Chalaza end (b) Egg cell
(c) Micropylar end (d) Suspensor

111. Assimilates in maize, are unloaded from the phloem terminals for accumulation in the endosperm at which one of the following?
(a) Funilcle (b) Micropyle
(c) Pedicel (d) Peduncle

112. In seed, generally assimilates reached from tissue to store by which one of the following?
(a) Diffusion (b) Osmosis
(c) Phloem bundles (d) Symplastically

113. In case of cytoplasmic-genetic male sterile system, the restorer genes are located in which of the following?
(a) Nucleus (b) Cytoplasm
(c) Both nucleus and cytoplasm
(d) Either nucleus or cytoplasm

114. Which one of the following terms is used for the radicle of cereals is covered by a protective sheath?
(a) Coleoptile (b) Coleorhiza
(c) Mycorhiza (d) Scutellum

115. Embryosac and integument have which one of the following?
(a) Connection of phloem strands
(b) Connection of xylem strands
(c) No symplastic connection
(d) Symplasitc connection

116. Mainly in which one of the following form, Carbon translocate into the seed?
(a) Amino acid (b) Fat
(c) Protein (d) Sucrose

117. The maintenance progeny of a genetic male sterile line are which of the following?
(a) 50% fertile and 50% sterile (b) Fertile
(c) Sterile (d) Not fixed

118. The process of germination in which cotyledon(s) remains beneath the soil is called as which one of the following?
(a) Epigeal (b) Hyogeal
(c) Hypo-epigeal (d) Viviparous

119. Mainly in which one of the following form, Nitrogen translocate into the seed?
(a) Amino acid (b) Fat (c) Protein (d) Sucrose

120. Reduce amylopectin content in the cotyledon of *Pissum sativum* is responsible for which one of the following?
(a) Green cotyledon (b) Smooth seed coat
(c) Wrinkled seed coat (d) Yellow cotyledon

121. All the offsprings of a cytoplasmic male sterile line are which of the following?
(a) 50% fertile and 50% sterile (b) Fertile
(c) Sterile (d) Not fixed

122. *Which one of the following form of amino acid is translocated from the parent plant?
(a) Aldehyde group (b) Amide containing group
(c) Hydroxyl group (d) Keto containing group

123. The 'switch' from a developmental to germinative mode is elicited by which one of the following in a normal seed?
(a) Bursting event (b) Hydration event
(c) Maturation drying event (d) Maturation event

12.3 Seed Chemistry

1. The crop barley, maize, oats, rye, wheat, castor bean, oil palm reserves food viz. carbohydrate, fat, protein etc. in which one of the following storage organs?
   (a) Cotyledons (b) Endosperm
   (c) Gametophyte (d) None of these
2. Which of the following do not provide energy to the seed required during germination and emergence?
   (a) Carbohydrates (b) Fats and Oils
   (c) Proteins (d) Vitamins
3. Which one of the following is the dominant crop for starch production?
   (a) Cassava (b) Maize
   (c) Potato (d) Sweet Potato
4. Starch may be easily broken down enzymatically to glucose monomers. *Amylose* forms of starch is stored in seeds. Which one of the following is not true about amylose?
   (a) 100% digestible by á-amylase.
   (b) Branched polymer formed by D-glucose units.
   (c) Linked by á-1, 4 glucosidic bonds.
   (d) Stains dark blue when exposed to iodine.
5. The crop broad bean, pea, peanut, soybean reserves food viz. carbohydrate, fat, protein etc. in which one of the following storage organs?
   (a) Cotyledons (b) Endosperm
   (c) Gametophyte (d) None of these
6.. Highest amounts of proteins are found in which one of the following leguminous crops?
   (a) Broad bean (b) Garden pea
   (c) Peanut (d) Soybean
7. Starch is an insoluble polysaccharide carbohydrate that is formed by the polymerization of many glucose moieties. In seeds, starch is stored in two forms in which *amylopectin* is one of them.
   (a) Linked by á-1, 4 and á-1, 6 glucosidic bonds.
   (b) Only 50% digestible by á-amylase.
   (c) Stains purplish-red when exposed to iodine.
   (d) Unbranched polymer formed by D-glucose units.

8. β-oxidation of fatty acids takes place at which one of the following part of seed's cell?
(a) Cell cytoplasm (b) Inner mitochondrial membrane
(c) Matrix of mitochondria (d) Ribosomes

9. Lowest amounts of carbohydrates are found in which one of the following leguminous crops?
(a) Broad bean (b) Garden pea
(c) Peanut (d) Soybean

10. Which one of the following is correct statement, except?
(a) Maize is the dominant crop for starch production.
(b) Seeds are rich in hemicelluloses, if starch is absent in seed.
(c) Most starch grains are composed of 50-75% amylopectin and 20-25% amylose.
(d) Starch is a soluble polysaccharide as amylose and amylopectin found in seed.

11. In seeds, lipid reserves are accumulated in discrete sub-cellular organelles known as oil-bodies. They are generally also known as, except which one of the following?
(a) Glyceriomes (b) Lipid bodies
(c) Oleosomes (d) Spherosomes

12. The liquid endosperm of coconut is rich in which one of the following growth promoter?
(a) Auxin (b) Cytokinins
(c) Ethylene (d) Gibberellins

13. Which of the following is not an unsaturated fatty acid?
(a) Palmitic acid (b) Oleic acid
(c) Linoleic acid (d) Erucic acid

14. Different plant species may have starch grains of characteristics shapes. Which one of the following is supporting to the statements?
(a) Angular in maize (*Zea mays*)
(b) Elliptical in runner bean (*Phaseolus coccineus*)
(c) Spherical in barley (*Hordeum vulgare*)
(d) All of the above

15. Which of the following are not main components of a seed are found in its food reserves?
(a) Carbohydrates (b) Fats and Oils
(c) Proteins (d) Vitamins

16. Much of the starch produced by maize is hydrolyzed to glucose, which is converted primarily into which one of the following?
(a) Fructose and Ethylene (b) Glucose and Ethylene
(c) Glucose and Fructose (d) None of these

17. Species from which one of the following family contribute more to human nutrition than any other plant family?
(a) Gramineae (b) Leguminosa
(c) Poaceae (d) Both (a) and (c)

18. The aleurone layer of cereal seed is rich in which one of the following?
(a) Carbohydrates (b) Fats
(c) Proteins (d) Vitamins

19. The principal site for lipid storage in cereal seeds is which one of the following?
(a) Cotyledons (b) Gametophyte
(c) Scutellum (d) None of these

20. Nutritional use of seed oils of species such as sunflower and maize has been favored by their high content of which one of the following fatty acids?
(a) Linoleic acids (b) Linolenic acid
(c) Oleic acids (d) Palmitoleic acid

21. In dycotyledonous seeds, lipids are stored primarily in which one of the following?
(a) Cotyledons (b) Endosperm
(c) Gametophyte (c) None of these

22. The shape of the starch grains is mainly determined by the __________ content, being rounder with higher ________content.
(a) Amylose (b) Amylopectin
(c) Amylose and Amylopectin (d) Amylopectin and Amylose

23. Which one of the following enzyme is different from the rest of others?
(a) Amylase (b) Lipase
(c) Maltase (d) Phosphorylase

24. In general, cereal seeds are rich in which one of the following proteins?
(a) Globulins (b) Glutelins
(c) Prolamins (d) Both (b) and (c)

25. Which of the following is not a factors that affect their quantity and quality in seeds?
(a) Carbohydrates (b) Fats and Oils
(c) Proteins (d) Vitamins

26. Less frequently, carbohydrates are stored as accumulations of which one of the following in the cell walls of the cotyledons?
(a) Arabinogalactan (b) Xyloglucans (amyloids)
(c) Both (a) and (b) (d) None of these

27. If starch is absent in seeds that are rich in hemicelluloses as the second most important form of carbohydrates stored in seeds which is generally composed of five sugars. Including xylose and arabinose, which one of the following is not among them?
(a) Fructose (b) Galactose
(c) Glucose (d) Mannose

28. Seeds of legumes and other dicotyledonous seeds have high concentrations of which one of the following proteins?
(a) Globulins (b) Glutelins
(c) Prolamins (d) Both (b) and (c)

29. *Starch grains* refer to the discrete subcellular bodies as starch found I\in the seed. Most of the starch grains are composed of which one of the following?
(a) 20-25% amylose
(b) 50-75% amylopectin
(c) 50-75% amylose and 20-25% amylopectin
(d) Both (a) and (b)

30. During the germination of monocot seeds, the hydrolytic enzymes are synthesized from which of the following parts of seed?
(a) Aleurone layer (b) Endosperm
(c) Scutellum (d) Seed coat

31. In maize seeds, zein (a prolamin) is the most abundant storage protein. Zein is relatively in which one of the following amino acids?
(a) Alanine (b) Leucine
(c) Tryptophan (d) Both (a) and (b)

32. Unsaturated fatty acids are the predominant found in oil seeds are
(a) Linoleic acids (b) Oleic acids
(c) Palmitoleic acid (d) Both (a) and (b)

33. The major storage protein in wheat is which one of the following, also abundant in rye and barley seeds?
(a) Gluten (b) Gliadin
(c) Glutenin (d) None of these

34. Match the following starches food sources with their appropriate amylose contents.
1. Corn A. 20
2. Wheat B. 18
3. Sweet Potato C. 17
4. Potato D. 26
5. Cassava E. 28
(a) 1E 2A 3D 4B 5C (c) 1E 2D 3A 4B 5C
(b) 1E 2D 3B 4A 5C (d) 1E 2B 3A 4D 5C

35. Match the following chemical compounds with their suitable seed as where they are found
1. Caffeine A. Red and Black beans
2. Hyoscyamine B. Tea seed
3. *Tannins* C. Horse nut
4. Saponin D. *Datura stramonium.*
5. Esculin E. *Coffea arabica* and *Coffea robusta*
(a) 1E 2A 3D 4B 5C (c) 1E 2D 3A 4B 5C
(b) 1E 2D 3B 4A 5C (d) 1E 2B 3A 4D 5C

12.4 Seed Health

1. A variety used by farmer was resistant to yellow mosaic virus (YMV) of soybean but in the current year the symptoms of disease appeared on the plants. It shows deterioration of which one of the following?
(a) Crop (b) Seed (c) Soil (d) Variety

2. Which of the following is the objectionable weed plant for wheat seed crop?
(a) *Convolvulus arvensis* (b) *Lathyrus* spp.
(c) *Phalaris minor* (d) *Vicia* spp.

3. Sample dawn from different bags of a seed lot is called as which one of following?
(a) Primary sample (b) Secondary sample
(c) Submitted sample (d) Working sample

4. Diaphanoscope is used to test which one of the following?
(a) Genetic purity (b) Moisture content
(c) Physical purity (d) Seed viability

5. For seed crop of wheat, which one of the following is considered as an objectionable disease?
(a) Blight (b) Leaf spot (c) Loose smut (d) Rust

6. Which one of the fungicidal formulation is more effective to kill the internal seed-borne pathogens?
(a) Dust (b) Liquid (E(C)
(c) Wettable powder (d) All of the above

7. In sorghum and pearl millet both, which one of the following disease is objectionable at seed level?
(a) Downy mildew (b) Ergot
(c) Grain smut (d) Head smut

8. For a seed crop of mustard, which of the following is specified as objectionable weed?
(a) *Argemone Mexicana* (b) *Brassica chinensis*
(c) *Brassica pekinensis* (d) *Phalaris minor*

9. Seed of pearl millet (from fields) having ergot infection even within the prescribed limits to become eligible for certification. Which one of the following solution should be subjected to floatation treatment?
(a) Brawn solution (b) Brine solution
(c) Brown solution (d) None of these

10. The seed certification standard of Karnal bunt for foundation class is 0.05%. It means one has to observe minimum how many seeds are from of the following?
(a) 100 (b) 1000 (c) 10000 (d) 100000

11. Among the plant diseases given below, which is not a seed borne disease for cole crops?
(a) Black leg (b) Black rot (c) Mosaic (d) Soft rot

12. Genetic purity is maintained by which one of following in maintenance breeding programme?
(a) Rejection of the line in the event all the plants of the line are off-type.
(b) Removal of the whole line in which off type plant appears.
(c) Rouging is never required in maintenance breeding.
(d) Rouging of offtype plant.

13. *Orobanche cumna* is a designated objectionable parasitic weed of which one of the following in India?
(a) Safflower (b) Soybean
(c) Sugarcane (d) Sunflower

14. For seed crop of cucumber which of the following is the objectionable weed plant?
    (a) *Cucumis hardwicki* (b) *Cucumis melo*
    (c) *Cucumis prophetarum* (d) All of the above

15. *Cuscutta spp.* is a designated objectionable parasitic weed of which one of the following in India?
    (a) Egyptin clover (b) Lucerne
    (c) Niger (d) Oat

16. Stripe rust wheat is due to which one of the following causal organism?
    (a) P, graniinis recondite (b) P. gralllillis strifornti s
    (c) P. gratuiuis tritici (d) None of these

17. Which one of the following is not observed during physical purity analysis by number?
    (a) Other crop seed (b) Weed seed
    (c) Objectionable weed seed (d) Other variety seed

18. Which of the following is the seed borne disease of tomato?
    (a) Early blight (b) Leaf spot
    (c) TMV (d) All of the above

19. Traps put in the store to catch the insects like *Tribolium* and *Sitophilus* are which one of following?
    (a) Pheromone (b) Allomone
    (c) Karomone (d) Hormone

20. Seed sample of maize and pulses with insect infestation are rejected during certification, if they are above one of the following?
    (a) 0.1% (b) 0.5% (c) 1.0% (d) 10%

21. For which of the following seed crops, chicory (*Cichorium intybus*) is specified as the objectionable weed?
    (a) Berseem (b) Cowpea
    (c) Groundnut (d) Lucerne

22. Seed sample other than maize and pulses with insect infestation are rejected during certification, if they are above one of the following?
    (a) 0.1% (b) 0.5% (c) 1.0% (d) 10%

23. Which of the following is considered as an objectionable weed for mustard seed crop?
    (a) *Argemone Mexicana* (b) *Brassica chinensis*
    (c) *Brassica pekinensi* (d) *Brassica tourneforti*

24. For which of the following seed crops, dodder (*Cuscuta* spp.) is specified as the objectionable weed?
(a) Berseem (b) Cowpea
(c) Groundnut (d) Lucerne

25. In barley, wheat, oat and triticale to avoid transmission of loose smut, the isolation distance should be maintained at which one of the following in certified seed production programme?
(a) 100m (b) 150m (c) 200m (d) 50m

26. Which one of the following crop considered to Top borer, Inter-node borer, Stalk borer, *Plassey* borer, *Gurdaspur* borer, Scale insect, Mealy bug etc. are the objectionable insect pest?
(a) Maize (b) Pearl millet
(c) Sorghum (d) Sugarcane

27. Primarily, which one of the following is detected by blotter method of seed health testing?
(a) Bacteria (b) Fungi
(c) Nematodes (d) Virus

28. Johnson grass is considered as a genetic contaminant for which of the following seed crop?
(a) Maize (b) Pearl millet
(c) Sorghum (d) None of these

29. Which one of the following plant disease is caused by seed borne nematode?
(a) Ear cockle of wheat (b) Ergot of rye
(c) Root knot of brinjal (d) None of these

30. If the field is infected by brown rot, wart or nematode, this field should not be selected for seed production of which one of the following crops?
(a) Onion (b) Potato tubers
(c) Soybean (d) Sugarcane

31. By number, the seed of black soybean present in yellow seeded variety of soybean in physical purity analysis would be called as which one of following?
(a) Inert matter (b) OCS
(c) ODV (d) Pure seed

32. Halo blight is the seed borne disease of which one of the following pulses crops?
(a) Mung bean (b) Rajmash
(c) Soybean (d) Urd bean

33. For which one of the following crops, *Cercospora* leaf spot is the objectionable seed borne disease?
(a) Ground nut (b) Linseed
(c) Niger (d) Sesame

34. For sunflower and pearl millet both crops, which one of the following disease is objectionable at field level?
(a) Cercospora leaf spot (b) Downy mildew
(c) Halo blight (d) Head smut

35. Which one of following term is used in physical purity analysis by weight immature, shriveled, diseased germinated or under sized seed of the crop under test?
(a) Inert matter (b) Pure
(c) ODV (d) Unhealthy seed

36. Which one of the following test is indicating the capacity of seed to form normal healthy seedlings under optimum conditions?
(a) Germination test (b) Seed vigour test
(c) Viability test (d) None of these

37. Seed production programme of cowpea, French bean, cluster bean and Indian bean should pass the seed certification standard for infection of only in hilly areas is which one of the following?
(a) Ascochyta leaf blight (b) Rhizoctonoa root rot
(c) Fusarium wilt (d) Yellow mosaic virus

38. Which of the following seed extraction method in tomato effectively controls bacterial canker disease?
(a) Acid extraction method (b) Alkali extraction method
(c) Fermentation method
(d) Mechanical seed extraction method

39. To raise a seed crop of potato, the land shall be free from infection of brown rot and cyst forming nematode for at least _______ years.
(a) 1 (b) 2 (c) 3 (d) 4

40. Both at seed and field level, which one of the following are an objectionable disease of pearl millet?
(a) Downy mildew (b) Ergot
(c) Grain smut (d) Head smut

41. Which one of the following is a common disease for wheat, triticale, oat, barley crops at field level?
(a) Black rust (b) False smut (c) Loose smut (d) Yellow Rust

42. In seed production of chickpea, which one of the following is an objectionable pest in India?
(a) Fusarium wilt (b) Helicoverpa
(c) Rhizoctonia root rot (d) None of these

43. Seed fields that can be re-inspected for which one of the following?
(a) Pearl millet infected by downy mildew
(b) Sorghum infected by ergot
(c) Sorghum infected by grain and head smut
(d) Wheat infected with Loose smut

44. Downy mildew of pearl millet is a serious seed borne disease at which one of the following?
(a) Field level (b) Seed level
(c) Both (a) and (b) (d) None of these

45. At seed level, which one of the following is a objectionable weed of rice in India?
(a) *O. nivara* (b) *O. rufipogon*
(c) *O. sativa* var. *fatua* (d) *Zizania palustris*

46. At seed level, which one of the following is a objectionable weed of rape seed and mustard in India?
(a) *Argimone mexicana* (b) Carthamus tinctorious
(c) *Cichorium intybus* (d) *Orobanche*

47. Which one of the following is the objectionable fungal disease of wheat both at field and seed level?
(a) Karnal bunt (b) Loose smut
(c) Rust (d) Blight

48. *Carthamus oxyacantha* is the objectionable weed of which one of the following crops?
(a) Linseed (b) Safflower
(c) Soybean (d) Sunflower

49. At seed level, which one of the following is a objectionable weed of Egyptian clover in India?
(a) *Chicorium rotudus* (b) *Cichorium intybus*
(c) *Cyanodon dactylus* (d) None of these

50. Which one of the following is the objectionable weed of wheat both at field and seed level?
(a) *Convolvulus arvensis* (b) *Cyprus rotandus*
(c) *Melilotus alba* (d) None of these

51. *Phalaris minor* is an objectionable weed of wheat at which one of the following levels?
(a) Field level (b) Seed level
(c) Both (a) and (b) (d) None of these

52. Cotton leaf roller is an important insect pest of which one of the following?
(a) Cucumber (b) Jute
(c) Okra (d) Sunflower

53. Infestation of granary weevil, rice weevil, grain borer, Angoumois grain moth, seed beetles etc. are
(a) No infestation in store (b) Not visible from the out side
(c) Only on the surface (d) Visible from the out side

54. Granary weevil, rice weevil, grain borer, Angoumois grain moth, seed beetles etc. lay their eggs in seeds
(a) At the time of harvesting
(b) At the time of processing
(c) When seeds are attached to the plant and are at milk stage
(d) None of these

55. Sodium hydroxide (NaOH) seed soak method is as approve by the Central Seed Certification Board, for testing which one of the following seed infections as per certifications standards for foundation seed and certified seed?
(a) Bunt of paddy (b) Ergot of pearl millet
(c) Kumal bunt of wheat and triticale (d) Both (a) and (c)

56. Visual observation method is as approve by the Central Seed Certification Board, for testing which one of the following seed infections as per certifications standards for foundation seed and certified seed?
(a) Ergot of pearl millet
(b) Most of the potato disease
(c) Soft and Basal rot of onion (d) All of the above

57. Introduction of new strains or physiologic races of the pathogen along with new germplasm from other countries may cause seed-born disease. The seed-borne pathogens may result due to which one of the following?
(a) Toxin substances production in infected seed
(b) Loss in germination, discoloration and shriveling
(c) Development of plant diseases and distribution of pathogen to new areas.
(d) All of the above

58. A pathogen may penetrate several of these parts of the seed and in turn infect them. The pathogen may be carried with the seeds in which one of the following ways?
(a) The pathogen may be present on seed surface.
(b) Pathogens are independent of seeds but accompany them.
(c) Pathogens establish within the seed with definite relationship with seed parts.
(d) All of the above

59. The pathogen may be present on seed surface as spores, oospores and chlamydospores as in case of which one of the following?
(a) Kamal bunt of wheat (b) Covered smut of barley
(c) Downy mildew of pear millet (d) All of the above

60. Rust disease of tea (*Camellia sinensis*) is caused by which one of following micro-organisms?
(a) Algae (b) Fungi (c) Virus (d) None of these

61. Embryo count method method is as approve by the Central Seed Certification Board, for testing which one of the following seed infections as per certifications standards for foundation seed and certified seed?
(a) Early blight of potato (b) Ergot of pearl millet
(c) Late blight of potato (d) Loose Smut of wheat

62. The seed primordium or the maturing seed infection from outside may be introduced through which one of following?
(a) Flower or Fruit stalk (b) Seed coat
(c) Stigma or Ovary wall or Pericarp (d) All of the above

63. The storage fungi *Aspergillus* and *Penicillium* can invade most types of seeds under which one of the following moisture storage conditions?
(a) Low (b) Medium
(c) High (d) None of these

64. The seed primordium or the maturing seed may be infected by directly from the infected plant through which one of following?
(a) Directly from the seed surface
(b) Infected plant through Flower
(c) Infected plant through Fruit and Seed stalk
(d) All of the above

65. In India, which one of the following is the objectionable seed borne disease of soybean in seed certification?
(a) Cercospora leaf spot (b) Root rot
(c) Yellow Mosaic Virus (d) None of these

66. Downey mildew is an objectionable seed borne disease for which one of the following crops?
(a) Sunflower (b) Pearl millet
(c) Both (a) and (b) (d) None of these

67. The continuous sowing of infected seed with which one of the following, the soil may become sick?
(a) Downey mildew of pearl millet and Halo blight of mung.
(b) Karnal bunt of wheat and downy mildew of sunflower.
(c) Loose smut of wheat and Cercospora leaf spot of sesame.
(d) Rice bunt and head smut of sorghum.

68. By number, physical purity analysis is performed on which one of following?
(a) Composite sample (b) Submitted sample
(c) Working sample (d) Secondary sample

69. Which one of the following is/are true statements?
(a) The fungi *Ergot sclerotia* are mixed with healthy seeds during threshing.
(b) The pathogen may be carried with the seeds in three ways; Admixture, External and Internal.
(c) The process of seed infection is influenced by the conditions under which the crop grows.
(d) All of the above

70. Which one of the following is not considered a component of seed health?
(a) Seed abnormality caused due to deficiency of trace elements
(b) Seed borne pathogens
(c) Storage pests of seeds
(d) None of these

71. The race – T of *Helminthosporium maydis* was introduced to India through seeds imported from which one of the following countries?
(a) Canada (b) UAS (c) UK (d) USSR

72. Which one of the following mycotoxin is produced in seeds infected by the fungus *Aspergillus flavus*?
(a) Aflatoxin (b) Nephrotoxin
(c) Ochratoxin (d) Rubratoxin

73. Which one of the following mycotoxin is produced in seeds infected by the fungus *Penicillium viridicum*?
(a) Aflatoxin (b) Nephrotoxin
(c) Ochratoxin (d) Rubratoxin

74. Which one of the following pathogen is exclusively seed-borne in nature?
(a) *Alternaria* spp (b) *Helminthosporium* spp
(c) *Pyricularia* spp (d) *Ustilago* spp

75. Which one of the following seed-borne pathogen is short lived in nature?
(a) *Ascochyta pisi* (b) *Claviceps purpurea*
(c) *Helminthosporium maydis* (d) *Ustilago nuda*

76. The incidence of viral disease in a seed field can be checked by application of which of these pesticide ?
(a) Bactericide (b) Fungicide
(c) Herbicide (d) Insecticide

77. Which one of the following are not a soil-borne pathogen affecting seeds and seedlings?
(a) *Aspergillus* spp (b) Fusarium spp
(c) Phytophthora spp (d) Pythium spp

78. Which one of the following plant disease can be controlled by mechanical separation of the disease causing agents from the seeds?
(a) Downy mildew of pearl millet (b) Ergot of rye
(c) Pod spot of pea (d) Smut of wheat

79. Which one of the following is not a bio-control agent used for seed treatment?
(a) *Bacillus subtilis* (b) *Penicillium notatum*
(c) *Pseudomonas fluroscens* (d) *Trichoderma viride*

80. Acid delinting of cotton seeds is effective in controlling which one of the following diseases?
(a) Angular leaf spot (b) Anthracnose
(c) Wilt (d) None of these

81. Which one of the following insect pest of stored seeds feeds only the embryo portion of seed?
(a) Anguimois grain moth (b) Indian meal moth
(c) Khapra beetle (d) Lesser grain borer

82. Which one of the following insect pest of stored seeds makes entry into the seed lot by damaging the container/bag?
(a) Cadelle (b) Khapra beetle
(c) Rice weevil (d) Saw toothed beetle

83. A seed abnormality in ground nut commonly called 'hollow heart of groundnut' is caused due to which one of the following reasons?
(a) Adverse environmental condition during seed development
(b) Deficiency of boron
(c) Deficiency of manganese
(d) Deficiency of potash

84. A seed abnormality in pea commonly called 'hollow heart of pea' is caused due to which one of the following reasons?
(a) Adverse environmental condition during seed development
(b) Deficiency of boron
(c) Deficiency of manganese
(d) Deficiency of potash

85. Brown spot of rice is caused by which one of the following pathogens?
(a) *Bipolaris oryzae* (b) *Dresclera oryzae*
(c) *Helminthosporium oryzae* (d) All of the above

86. The level of which one of following enzyme's activities determine to the tetrazoiium test?
(a) ATPase (b) Carboxylase
(c) Dehydrogenase (d) Isomerase

87. Match the following disease of crops with their causal organism.

1. Rust (red) of tea	A. *Claviceps fusiformis* = *Claviceps microcephela*
2. Kumal bunt of wheat and triticale	B. *Neovossia horrida*
3. Bunt of paddy	C. *Sphacelia sorghi* and *Claviceps* spp.
4. Ergot of pearl millet	D. *Neovossia indica*
5. Ergot or sorghum	E. *Cephaleuros virescens* = *Cephaieuros parasiticus*

(a) 1E 2A 3D 4B 5C (c) 1E 2D 3A 4B 5C
(b) 1E 2D 3B 4A 5C (d) 1E 2B 3A 4D 5C

88. Match the following important disease of potato crops with their causal organism.

1. Late blight	A. *Alternaria solani*
2. Early blight	B. *Streptomyces scabies*
3. Black scrul	C. *Synchytrium endobioticum*
4. Common scab	D. *Rhizoctonia solani*
5. Wart	E. *Phytophthora infestans*

(a) 1E 2A 3D 4B 5C (c) 1E 2D 3A 4B 5C
(b) 1E 2D 3B 4A 5C (d) 1E 2B 3A 4D 5C

89. Match the following rot of potato crops with their causal organism.

1. Dry rot	A. *Sclerotium rolfsii*
2. Charcoal rot	B. *Phytophthora erythroseptica*
3. Wet rot of potato	C. *Erwinia* spp.
4. Pink rot	D. *Macrophomina phaseolina*
5. Blackleg and Soft rot	E. *Fusarium coeruleum*

(a) 1E 2A 3D 4B 5C (c) 1E 2D 3A 4B 5C
(b) 1E 2D 3B 4A 5C (d) 1E 2B 3A 4D 5C

90. Match the following disease of crops with their causal organism.

1. Loose Smut of wheat.	A. *Erwinia carotovora*
2. Ear cockle of wheat	B. *Anguina tritici*
3. Soft rot of onion	C. *Pseudomonas aeruginosa*
4. Basal rot of Onion	D. *Fusarium oxysporum* f. cepae
5. Brown rot of onion	E. *Cstilago tritici*

(a) 1E 2A 3D 4B 5C (c) 1E 2D 3A 4B 5C
(b) 1E 2D 3B 4A 5C (d) 1E 2B 3A 4D 5C

12.5 Seed Production Techniques

1. Among the 123 hybrid seed producer of cotton, I2 have submitted seed of female as hybrid. It may be verified with the help of test at field level.
   (a) DUS (b) GOT (c) ODV (d) VCU

2. The compact area approach of seed production is advocated under the seed village scheme for which of the following reasons?
   (a) Avoids isolation and rouging problems.
   (b) Facilitates seed certification.
   (c) It is easy to provide technical guidance, supervision and training to the seed growers.
   (d) All of the above

3. India has a well-organized seed production programme. Which one of the following organizations is responsible for seed multiplication, certification and distribution of high quality seed?
   (a) National Seeds Corporation (NS(C)
   (b) State Seed Certification Agencies (SSCAS)
   (c) State Seeds Corporations (SSCS)
   (d) All of the above

4. Genetic deterioration through natural crossing is not at all a problem in which of the following group of crops?
   (a) Cross pollinated crops (b) Often-cross pollinated crops
   (c) Self-pollinated crops (d) Vegetatively propagated crops

5. For production of certified seed of hybrid category with the use of three line breeding, foundation seed would be which one of following?
   (a) Seed of A and B line (b) Seed of A and R line
   (c) Seed of B and R line (d) None of these

6. According to Bateman (1942) the extent of genetic contamination in the seed field through natural crossing depends on?
   (a) Isolation distance (b) Pollinating agents
   (c) Varietal mass (d) All of the above

7. Which one of the following purity is maintained by avoiding Out crossing and mixture of seeds of other varieties in the produced seed lot?
   (a) Genetic (b) Health
   (c) Physical (d) None of these

8. Before growing of the seed crop, the cropping history of the seed field is seen to check which one of the following?
(a) Diseased plants (b) Other crop plants
(c) Volunteer plants (d) Weed plants

9. For hybrid seed production of pearl millet, which one of following male and femalke row ratios would be?
(a) Equal number of male and female lines.
(b) Less number of female and more number of male lines.
(c) More number of female and less number of male lines.
(d) No ratio between male and female line is required.

10. Which one of the following is different from each other?
(a) Isolation distance (b) Space isolation
(c) Spatial isolation (d) Specific isolation

11. Which one of the following is the purity of the seed that is improved by rouging?
(a) Genetic purity (b) Physical purity
(c) Ethic purity (d) Expressed purity

12. Which of the following is not a synonym of others?
(a) Barrier isolation (b) Differential blooming
(c) Temporal isolation (d) Time isolation

13. Which one of following requires for hybrid seed production involving three line breeding?
(a) More seed of A than B line (b) More seed of A than R line
(c) More seed of B than A line (d) More seed of R than A line

14. An important operation during seed production which is generally termed as negative mass selection would be which one of the following?
(a) Field inspection (b) Isolation
(c) Rouging (d) Synchronization

15. By which one of the following, the genetic impurity in pigeonpea due to out crossing can be sorted out?
(a) DUS test (b) GOT test
(c) ODV test (d) Both (b) and (c)

16. For combine harvesting of seed crop of paddy, the seed moisture content should not be exceed?
(a) 18 (b) 20 (c) 23 (d) 25

17. For cultivation of homogeneous population of heterozygous plants, certified seed of which one of following would be provided to farmers?
(a) Composite (b) Hybrid (c) Synthetic (d) None of these

18. **The environmental sensitive genetic -male sterility is usually exploited for hybrid seed production in which one of the following crops?**
(a) Maize (b) Pearl millet
(c) Rice (d) Sorghum

19. **During seed production of wheat, the designated inseparable crop plants are which one of following?**
(a) Chickpea (b) Lentil
(c) Linseed (d) Mungbean

20. **Which one of the following treatment is made for advancing flowering of late parental line in rice hybrid seed production?**
(a) Application of GA_3
(b) Application of KH_2PO_4
(c) Application of quick releasing nitrogenous fertilizer
(d) None of these

21. **Which one of following is considered for commercial hybrid seed production of maize?**
(a) Spadix is removed form female parent.
(b) Spadix is removed form male parent.
(c) Tassel is removed from female parent.
(d) Tassel is removed from male parent.

22. **Which one of the following chemical is commonly used as a gelling agent for production of synthetic seeds?**
(a) Agar (b) Methyl cellulose
(c) Sodium alginate (d) Sodium pectate

23. **Maintenance breeding usually aims at production of which one of the following?**
(a) Foundation seed & Certified seed
(b) Nucleus seed & Breeder seed
(c) Truthfully Labeled Seed
(d) None of these

24. **Which of the following diseases of rice is caused by seed borne nematodes?**
(a) Ufra/Dakpora (b) White tip
(c) Both (a) and (b) (d) None of these

25. **The ideal planting ratio of female and male parental lines for hybrid sorghum seed production is which one of the following?**
(a) 4:1 (b) 4:2 (c) 6:1 (d) 6:2

26. At head formation stage a cross shape cut is made for seed stalk emergence in which one of following crops?
(a) Egyptian clover (b) Potato
(c) Cabbage (d) Castor

27. In wheat seed production, a minimum isolation distance of 150m is maintained to check which one of the following?
(a) Contamination of loose smut disease
(b) Control of rust disease
(c) Natural crossing
(d) Both a0 and (c)

28. Nucleus seed of carrot is produced by which one of following methods?
(a) Root to seed method (b) Seed to root method
(c) Seed to seed method (d) None of these

29. The method of synchronization by jerking of ear heads to delay flowering in early parent is followed in which one of the following?
(a) Maize (b) Pearl millet
(c) Rice (d) Sorghum

30. Which one of following method is used in commercial hybrid seed production programme of cotton that treat by hand emasculation and pollination involving male fertile female line?
(a) Doak method
(b) Self-incompatibility method
(c) Three line breeding method
(d) Two line breeding method

31. In the presence of artificial drying facility, it is advised to harvest the maize ears at a moisture content of which one of the following?
(a) 20-25% (b) 30-35% (c) 8-10% (d) 10-12%

32. Due to deterioration in which one of following, a farmer has to change the hybrid seed after every year?
(a) Genetic purity (b) Germinability
(c) Physical purity (d) Vigour

33. The minimum isolation requirements for production of FS and CS in pigeon pea should be which one of the following?
(a) 100 and 50 m (b) 200 and 100 m
(c) 300 and 150 m (d) 50 and 25 m

34. For hybrid seed production programme, deatsseling is required in which one of following crops?
(a) Maize (b) Pearl millet
(c) Rice (d) Sorghum

35. Threshing of cowpea pods for seed purpose by use of thresher should be done at a drum speed of ______ rpm.
(a) 100-200 (b) 200-300 (c) 300-400 (d) 400-500

36. Through proforma which one of following, the indent of breeder's seed production is allocated to different institutions?
(a) BSP I (b) BSP II (c) BSP III (d) BSP IV

37. The harvestable maturity stage of groundnut seed crop is indicated by which one of the following?
(a) Leaves start yellowing and fall down
(b) Pods become reticulated
(c) Seeds are separated from pods
(d) All of the above

38. The minimum isolation distance for Foundation Seed and Certified Seed production of sesame respectively, should be which one of the following?
(a) 100 and 50 m (b) 200 and 100 m
(c) 400 and 200 m (d) 50 and 25 m

39. The seed of maintainer line is maintained by which one of following in three line seed production programme?
(a) A XA (b) AXB (c) BXB (d) RXR

40. Nipping of the auxiliary buds of the plants to enhance seed yield is practiced in which one of the following crops?
(a) Castor (b) Niger
(c) Sunflower (d) Both (a) and (b)

41. Seed plot is grown at a particular distance from the sources of genetic contamination to avoid which one of the following?
(a) Cross pollination (b) Goitenogamy
(c) Out crossing (d) Self-pollination

42. The double cross hybrid seed production involves ______ number of inbred lines.
(a) 2 (b) 3 (c) 4 (d) 6

43. Dehaulming operation is related to seed production in which of the following seed crop?
(a) Groundnut (b) Onion
(c) Potato (d) Sweet potato

44. The optimum planting ratio for sunflower hybrid seed production is which one of the following?
(a) 2:1 (b) 3:1 (c) 4:2 (d) 6:1

45. In cytoplsmic and cytoplasmic genetic male sterility system, which can consider for the difference between maintainer and male sterile line?
(a) Cytoplasm (b) Nuclear gene
(c) Both (a) and (b) (d) None of these

46. In cotton seed production, higher yield of good quality seeds are obtained from bolls of which one among the following?
(a) Early pickings (b) Late pickings
(c) Mid pickings (d) All of the above

47. The most appropriate stage is which one of following for inspections of loose-smut-susceptible wheat's and cross-pollinated crops?
(a) Flowering (b) Maturity
(c) Post flowering (d) Pre flowering

48. Ginning of cotton seeds should be done at a seed moisture content of which one of the following percentage for maintaining quality of seed.
(a) 10.0-12.0 (b) 6.0-8.5 (c) 8.0-10.0 (d) 8.5-10.5

49. Certified seed produced from which one of following seeds?
(a) Breeder seed (b) Foundation seed
(c) Registered seed (d) All of the above

50. Delinted cotton seeds are considered superior to non-delinted seeds for the following reasons, except:
(a) Enables proper cleaning, grading and treatment
(b) Enhance germination rate
(c) Longer storability
(d) Seeds have good look

51. For hybrid seed production of which one of following crops cytoplasmic male sterility may be exploited?
(a) Cotton (b) Okra
(c) Potato (d) Soybean

52. Which one of the following techniques is followed for hybrid seed production in cotton?
(a) Hand pollination (b) Use of CHA
(c) Use of male sterility (d) Both (a) and (b)

53. Which one of following chemical is used for induction of male sterility?
(a) Abscisic acid (AB(A)
(b) Chemical hybridizing agents (CH(A)
(c) Hydroxycitric acid (HC(A)
(d) None of these

54. To enhance seed yield in jute, which one of the following operation would be useful?
(a) Clipping of apical buds (b) Nipping of auxiliary buds
(c) Spraying of gibberellic acid (d) None of these

55. During seed production programme disease escape accomplished by the avoidance of insect vector is termed as which one of the following term?
(a) Avoidance (b) Klenducity
(c) Tripping (d) Vector carnage

56. For induction of reproductive growth phase in sugar beet, a cool temperature below _____ °C for around two months is necessary.
(a) 7.5 (b) 10.5 (c) 15.5 (d) 20.0

57. The crops (viz. sorghum, pigeon pea, cotton, linseed, sesame require more isolation distance during seed production) categorized under which one of the following categories?
(a) Cross-pollinated (b) Often cross-pollinated
(c) Often self-pollinated (d) Self-pollinated

58. Which one of the following planting methods in sugarcane is generally considered ideal for production of better canes?
(a) Flat planting (b) Furrow planting
(c) Space transplanting (d) All of the above

59. In hybrid seed production programme, the restorer parent may have which one of following based on CGMS system?
(a) Heterozygous fertility gene on chromosome with fertility gene in cytoplasm.
(b) Homozygous dominant fertility gene on chromosome with sterility gene in cytoplasm.
(c) Homozygous recessive sterility gene on chromosome with fertility gene in cytoplasm.
(d) Homozygous recessive sterility gene on chromosome with sterility gene in cytoplasm.

60. The minimum isolation requirements for Certified Seed production in brinjal should be which one of the following?
(a) 100 m (b) 200 m
(c) 300 m (d) 50 m

61. Due to insect's pressure during nectar or pollen collection, the exposure of male and female part of flower is termed as which one of the following terms?
(a) Anthesis (b) Bursting
(c) Hammering (d) Tripping

62. Which one of the following treatment induces more number of female flowers in cucumber?
(a) High temperature (b) Longer photo period
(c) Low concentration of GA_3 (d) Low fertility

63. Due to which one of following, the sterility is expressed in CGMS system?
(a) Gene for sterility is present both in cytoplasm and nucleus.
(b) Gene for sterility is present in cytoplasm.
(c) Gene for sterility is present in nucleus or in cytoplasm.
(d) Gene for sterility is present in nucleus.

64 Which one of the following techniques is followed for hybrid seed production in cucumber?
(a) Hand pollination (b) Use of CMS line
(c) Use of gyno monoecious line (d) None of these

65. To induce seed stalk formation horizontal cut is made on which one of following crops?
(a) Curd of cauliflower (b) Potato tuber
(c) Sugarcane sett (d) Sweet potato

66. The optimum monthly average temperature for cauliflower seed production is ______ °C:
(a) 15-20 (b) 20-25 (c) 5-10 (d) 10-15

67. Buff colour tag is used for which one of following seeds?
(a) Breeder seed (b) Certified seed
(c) Foundation seed (d) Nuclear seed

68. Which one of the following method of cabbage seed production requires heavy staking?
(a) Head intact method
(b) Stump method
(c) Stump with central core intact method
(d) All of the above

69. During hybrid seed production programme, the male sterility system with nearly 50% male fertile plants in female line is due to which one of the following?
(a) CGMS (b) CMS (c) GMS (d) Self-incapability

70. Seed production of which one of the following cole crops is possible in plains?
(a) Cabbage (b) Cauliflower
(c) Chinese cabbage (d) Knolkhol

71. In hybrid seed production programme of which one among the following crops, space isolation can be altered?
(a) Maize (b) Pearl millet
(c) Sorghum (d) Sunflower

72. Root to seed method is preferred over seed to seed method of seed production in radish, garden beet and carrot for which one of the following advantage?
(a) Genetic purity is ensured (b) Seed yield is higher
(c) Takes comparatively less time (d) All of the above

73. In commercial hybrid seed production of which one of following crops supplementary pollination is a pre-requisite?
(a) Pearl millet (b) Pigeon pea
(c) Rice (d) Sorghum

74. In onion, bulb to seed method is preferred to seed to seed method, because of the following advantages:
(a) Genetic purity is ensured through selection of true to type bulbs
(b) Permits selection of healthy tubers
(c) Seed yield is very high
(d) All of the above

75. The pollen grains are tested for viability test by using which one of the following solution?
(a) 2% aceto-carmine (b) Agar-Agar
(c) Commusive blue (d) Iodine solution

76. In potato, 'Hill Seed' is produced at locations _______ m above mean sea level.
(a) 1000 (b) 1500 (c) 2000 (d) 2500

77. Multiple heads are generally found in which one of following line for hybrid seed production of sunflower?
(a) A line (b) B line
(c) R line (d) None of these

78. In absence of cold storage facilities, the sprouting of seed tubers can be suppressed by treatment of tubers with which one of the following?
(a) Fusarex (b) MH
(c) Thiourea (d) None of these

79. Which one of the following term is used for the act of removing the plants of the same species, which is deviating from the normal expression of the variety in a seed production plot?
(a) Cropping (b) Off-type removal
(c) Rouging (d) Weeding

80. Dehaulming operation to prevent the aphid infestation in seed crop pf potato in done hy which of the following methods?
(a) Cutting of plants from the base
(b) Spraying of 2% CuSO4 solution
(c) Spraying of 2.4-D
(d) Both (a) and (b)

81. Lines A and Lines B are referred as which one of following in terms of genetics?
(a) Isogonic lines (b) Male fertile lines
(c) Male sterile lines (d) None of these

82. For maintenance of the gynomoecious lines of cucumber, the plants are sprayed with which one of the following chemical to induce formation of male flowers?
(a) Ethrel (b) Gibberellic acid
(c) Potassium nitrate (d) Silver nitrate

83. The hybrid seed production programme based on CGMS should be kept in isolation to avoid which one of following?
(a) Cross pollination (b) Goitenogamy
(c) Out crossing (d) Self-pollination

84. For one hectare, which one of following is the seed rate of true potato seed?
(a) 2.5 t/ha (b) 1 kg/ha
(c) 100 kg/ha (d) 100 g/ha

85. Which one of following the line that should not be made available to others for total control on the commerce of the CGMS based hybrid?
(a) A line (b) B line
(c) I line (d) R line

86. Selective removal of undesirable plants of the seed crop on the basis of visual inspection in the field to improve quality of seed is known as

(a) Cropping (b) Off-type removal
(c) Rouging (d) Weeding

87. In hybrid seed multiplication chain, seed of 'A' and 'R' lines are considered as which one of following type of seed?

(a) Breeder seed (b) Certified seed
(c) Foundation seed (d) Parental seed

88. An isolation distance of three meter is recommended in many self pollinated crops to check which one of the following?

(a) Cross pollination (b) Infection of disease
(c) Mechanical mixture (d) Out crossing

89. Which one of the following is common among the crops: rice, maize, pearl millet, sunflower, safflower, castor, pigeon pea and cotton?

(a) Exalbuminous seed (b) Allogamy
(c) Commercial hybrid (d) Photo insensitivity

90. Which one among the following term is used for synchronization of the receptivity of the male sterile plant to the maximum pollen load of the pollinator for cross-pollination in hybrid seed production?

(a) Confounding (b) Nicking
(c) Staggering (d) Synchronous maturity

91. In determination of isolation distance, the marker character is used as tester of population with which one of the following type of genes?

(a) Additive (b) Dominant
(c) Epistitc (d) Recessive

92. R line is known as Restorer because considered as which one of following?

(a) It restores hybrid vigour (b) It restores male fertility
(c) It restores male sterility (d) Its use is restricted

93. During hybrid seed production programme, physical enhancement of pollination is termed as which one of following?

(a) Auxiliary pollination
(b) Complementary pollination
(c) Secondary pollination
(d) Supplementary pollination

94. Careful and systematic evaluation of a seed production field and the removal of all undesirable plants of the same crop is termed as which one of the following terms?
(a) Cropping (b) Off-type removal
(c) Rouging (d) Weeding

95. The highest seed replacement rate is of which one of following?
(a) Synthetic (b) Composite
(c) Hybrid (d) None of these

96. Which one of following is considered for maintaining a wheat variety?
(a) Single ear selection from Breeder seed production plot
(b) Single ear selection from ear to row progeny of nucleus seed production plot
(c) Single plant selection from Breeder seed production plot
(d) Single plant selection from single plant progeny of nucleus seed production plot

97. By CGMS system, the cytoplasm of a hybrid produced would always be of which one among the following?
(a) A line (b) B line
(c) R line (d) None of these

98. From the seed production plot of wheat, the removal of plant of wheat infested by loose smut is called as which one of the following?
(a) Cropping (b) Off-type removal
(c) Rouging (d) Weeding

99. Male sterile pollens are produced in three line breeding programme in which one of following line plants?
(a) A line plant (b) B line plant
(c) R line pant (d) None of these

100. In the cytoplasm of R Line which one of following gene is present?
(a) Male sterile (b) Male fertile
(c) No gene of sterility or fertility (d) Both (a) and (b)

101. In the seed production, isolation and roguing are two major field operations that differs from commercial cultivation to maintain which one of the following?
(a) Genetic purity and Seed health (b) Out crossing and Vigour
(c) Physical purity and Germination
(d) Uniformity and Stability

102. By which one of following may be produced commercially to the hybrid seeds in the event of unavailability of male sterility in self-pollinated field crops with small flower
(a) Doak method (b) Gametocide
(c) Rope pulling (d) Tripping

103. In a seed production plot, plant similar for distinguishing character(s) but deviates in the agronomical trait(s) like, days to flowering, plant height, disease reaction and days to maturity is designated as which one of the following?
(a) ODV (b) Off-type (c) Rogue (d) Volunteer plant

104. In which one of following system hybrid seed is produced by making cross between with heterozygous gene for sterility in pollen parent and homozygous gene for sterility in female parent?
(a) CGMS (b) CMS
(c) GMS (d) Self-incompatibility

105. A plant in the seed crop, which deviates from the norm for the cultivar, but does not obviously belong to another cultivar is refers by which one of the following?
(a) ODV (b) Off-type
(c) Rogue (d) Volunteer plant

106. Plant showing variation in expression of distinguishing characters or abnormal performance of the plant in comparison to normal expression of the cultivar i.e., plant of other cultivar, other species or diseased plant of the same cultivar is considered as which one of the following?
(a) ODV (b) Off-type
(c) Rogue (d) Volunteer plant

107. In a crop with availability of all the three types of male sterility system for which one of following, cost of hybrid seed production would be the highest?
(a) CGMS system (b) CMS system
(c) GMS system (d) None of these

108. In the seed production, knowledge of which one of the following type of characters of the variety is the pre-requisite for the persons engaged in rouging?
(a) Diagnostic and Phenological traits
(b) Expression against biotic stresses
(c) Polygenic and Oligogenic traits
(d) Qualitative and Quantitative traits

109. Theoretically rouging should be performed at which one of the following stage?
(a) After flowering (b) At the time of flowering
(c) At the time of maturity (d) Before flowering

110. When days to flowering in A and R line are same then which one of following has to be adopted for hybrid seed production of rice?
(a) Rope pulling (b) Seed treatment
(c) Spray of GA_3 (d) Staggered sowing

111. Testing of genetic purity at seed level is called as which one of the following?
(a) DUS (b) GOT (c) ODV (d) VCU

112. On which one of following Chemical Hybridizing Agents are applied during hybrid seed production programme?
(a) Male (b) Female
(c) Both (a) and (b) (d) None of these

113. Deterioration of a variety when the proportion of different states of unnoticed traits may reach to equilibrium by natural selection and express in new environment is known as
(a) Genetic drift (b) Genetic shift
(c) Natural selection (d) Residual Segregation

114. At plant level, testing of genetic purity is called as which one of the following?
(a) DUS (b) GOT (c) ODV (d) VCU

115. The staggering is provided at which one of following time for rice hybrid seed production programme?
(a) Time of flowering (b) Time of harvesting
(c) Time of nursery raising (d) Time of transplanting

116. In the field, plants of the same crop grown due to shattered seeds of previous season crop/variety along with the present season crop is termed as which one of the following?
(a) Assistant plant (b) Pollen shedder
(c) Volunteer plant (d) None of these

117. Undesirable natural out-crossing has which one of the following?
(a) No effect on morphology of the produced seed.
(b) Seed will be of larger seed size.
(c) Seed will be or darker in colour.
(d) Significant alteration on morphology of produced seed.

118. Hybrid seed produced in three line system express male fertility due to which one of following?
(a) Heterozygous condition for Fertility with fertile cytoplasm.
(b) Heterozygous condition for Fertility with sterile cytoplasm.
(c) Homozygous dominant condition for fertility with sterile cytoplasm.
(d) Homozygous recessive condition for sterility with fertile cytoplasm.

119. Potato tuber growth is stopped at what temperature (?) from the following?
(a) 40-42 (b) 30-32 (c) 35-37 (d) 38-40

120. Hybrid seed is harvested from which one of following in CGMS system?
(a) A line (b) B line (c) H line (d) R line

121. During seed multiplication, which one of the following cause of variation that cannot be controlled by seed producer?
(a) Mutation (b) Out crossing
(c) Recombination (d) Residual segregation

122. In a hybrid seed production programme involving CGMS system, very high fruit setting in any one plant of A line indicates the possibility of which one of following?
(a) Heterosis (b) Hybridity
(c) Male fertility (d) Male sterility

123. In general, which one of following is the smallest number of plants of one cross-pollinated variety that should be grown to ensure genetic integrity?
(a) 100 plants (b) 200 plants
(c) 500 plants (d) 1000 plants

12.6 Seed Processing

1. Which one of the following term is used for top most screens with larger holes than the desirable seed size to remove the inert matter of larger size?
(a) Aspirator (b) Grader
(c) Huller (d) Scalper

2. Seed processing plants belongs to which one of the following in public sector?
(a) Farmer (b) Private agency
(c) Seed certification agency (d) Seed Corporation

3. Gravity separator separates seeds and contaminating materials on the basis of difference in
(a) Size (b) Weight
(c) Surface texture (d) None of these
4. A specified quantity of processed seed of a variety and class produced by a grower is called as which one of the following?
(a) Certified seed (b) Farmers seed
(c) Processed seed (d) Seed lot
5. Which one of the following is a moisture resistant container?
(a) Aluminum laminated pouches
(b) Polyethylene bags (700 gauge)
(c) Sealed tins
(d) Ungasketed metallic bin
6. By removal of undesirable material, improvement in physical purity of seed lot and upgrading of seed quality through removal of damaged and undersized seed by mechanical devices with highest efficiency including minimum loss and damage to seed is called as
(a) Seed halogenation (b) Seed invigoration
(c) Seed processing (d) Seed treatment
7. Seeds differing in their surface texture can be separated by which of the following equipment?
(a) Inclined drapper (b) Magnetic separator
(c) Roll mill (d) All of the above
8. The major steps viz. pre-conditioning, basic cleaning and grading are of which one of the following?
(a) Grading (b) Quality control
(c) Seed processing (d) Seed treatment
9. The operation that prepares a seed lot for basic cleaning is known as which one of the following?
(a) Grading (b) Pre-cleaning
(c) Pre-conditioning (d) Pre-taming
10. Which one of the following processing plant layout is widely followed in our country?
(a) Combined design (b) Multi storey design
(c) Single storey design (d) All of the above
11. Pneumatic separator and aspirator separate seeds and contaminating materials on the basis of difference in
(a) Gravity (b) Terminal velocity
(c) Weight (d) All of the above

12. For removal of corn seeds from its cob, which one of the following equipment is used?
(a) Debearder (b) Huller
(c) Scarifier (d) Sheller

13. Which one of the following equipment removes tightly fixed husk from seeds of grasses to facilitate in the process of sowing and germination?
(a) Debearder (b) Huller
(c) Scarifier (d) Sheller

14. Give the right sequence of the following seed processing operations:
(i) Treating (ii) Cleaning
(iii) Bagging (iv) Drying
(a) ii-iii-i-iv (b) iii-iv-i-ii
(c) iv-iii-i-ii (d) iv-ii-i-iii

15. Which one of the following equipment scratches the hard seed coat to improve the process of germination by increasing exchange of water and oxygen in crops like lucerne, fababean, rice bean etc?
(a) Scalper (b) Scarifier
(c) Sheller (d) Stratifier

16. Which one of the following parameters is not indicated in the psychrometric chart?
(a) Dew point (b) Dry-bulb temperature
(c) Light intensity (d) Relative humidity

17. Which one of the following scientists developed to the thumb rules for seed storage?
(a) Douglos (b) E. H. Roberts
(c) H. F. Harrington (d) Thompson

18. In which one of the following seed processing, Huller, Sheller, Debearder, decorticator are the equipment are involved?
(a) Basic cleaning (b) Cleaning
(c) Grading (d) Pre-conditioning

19. The spiral separator separates seeds differing in their
(a) Shape (b) Size
(c) Surface texture (d) Weight

20. A vital link between production and marketing of seeds is termed as which one of the following?
(a) Colouring (b) Numbering
(c) Processing (d) Tanning

21. Seed of groundnut is stored as which one of the following?
(a) Dehulled seed (b) Kernel
(c) Nut (d) Seed

22. During forced air drying of seeds, the static pressure increases with which one of the following?
(a) Decrease in seed size
(b) Increase in seed depth
(c) Increase in seed moisture content
(d) All of the above

23. The step of seed processing that removes the larger, smaller, lighter and thicker adulterants as compared to the crop seed, from the seed lot is known as which one of the following?
(a) Basic cleaning (b) Grading
(c) Pre-conditioning (d) Separation

24. Top most screen of a seed cleaner/grader with larger hole than the desirable seed size to remove the inert matter of larger size than the seed is known as which one of the following?
(a) Aspirator (b) Debearder
(c) Grader (d) Scalper

25. In a well-designed seed dryer it is advisable to keep the static pressure below ____ inches.
(a) 0.5 (b) 1.0 (c) 2.0 (d) 3.0

26. A specified quantity of processed seed of a variety and class produced by a grower is called as which one of following?
(a) Seed lot (b) Seed batch
(c) Seed bunch (d) Seed cluster

27. The separator that separates the inert matter, other crop seeds, weed seeds and shriveled seeds from healthy leguminous seeds using the ability of a seed to roll due to its shape is known as
(a) Disc (b) Gravity
(c) Indent (d) Spiral

28. Disc separator separates seeds and contaminating materials on the basis of difference in which one of the following?
(a) Colour (b) Length
(c) Shape (d) Weight

29. By turning out the seed mass, which one of the following term is used for revolving separator that provides each seed a chance to fit into the indent?
(a) Disc separator (b) Gravity separator
(c) Indent cylindrical separator (d) Spiral separator

30. A separator consists of a stratifying deck surface mounted at a slight angle that reciprocates and tosses heavy seed in uphill direction, whereas, light material remains at downhill place is called as which one of the following?
(a) Disc separator (b) Gravity separator
(c) Indent cylindrical separator (d) Spiral separator

31. In heated air drying of seed, the temperature of the drying air should not exceed _____ °F:
(a) 100 (b) 110 (c) 115 (d) 120

32. The seed lot from which one of the following, samples are drawn should be relatively?
(a) Heterogeneous (b) Homogeneous
(c) Uniform (d) Variable

33. By which one of the following process, lighter inert matter and adulterant than the crop seeds are removed from the seed lot?
(a) Aspiration (b) Dehulling
(c) Hulling (d) Pre-conditioning

34. In hot air oven method, moisture content of the seed is calculated by which one of the following formula?

(a) $\frac{\text{Intial weight-Final weight}}{\text{Final Weight}} \times 100$

(b) $\frac{\text{Final Weight-Intial weight}}{\text{Intial Weight}} \times 100$

(c) $\frac{\text{Wet Weight}}{\text{Intial weight}} \times 100$

(d) None of these

35. If the depth of seed in a dryer is doubled, the static pressure would be which one of the following?
(a) Doubled (b) Quadrupled
(c) Remain same (d) Tripled

36. Which one of the following presented to the mash size of a screen?
(a) N/16 inch (b) N/4 inch
(c) N/64 inch (d) N/8 inch

37. Which one of the following is generally, the shape of the mash of top screen?
(a) Oblong (b) Round (c) Triangular (d) Wire mash

38. Eradication of fungal pathogens that have established within the seed tissues is termed as:
(a) Seed contamination (b) Seed disinfection
(c) Seed disinfestations (d) Seed protection

39. For safe seed storage, the total of relative humidity (in percent and temperature) in Fahrenheit should not exceed_________.
(a) 50 (b) 70 (c) 100 (d) 120

40. With the help of which one of the following, separation is done based on seed weight?
(a) Aspirator (b) Disc separator
(c) Gravity separator (d) Spiral separator

41. Which one of the following is different in respect of principle of operation?
(a) Cylinder separator
(b) Cylindrical screen separator
(c) Horizontal flat screen separator
(d) Vertical ribbed screen separator

42. With the help of which one of the following, separation is done based on seed shape?
(a) Aspirator (b) Disc separator
(c) Gravity separator (d) Spiral separator

43. With the help of which one of the following, separation is done based on seed surface texture?
(a) Aspirator (b) Disc separator
(c) Gravity separator (d) Spiral separator

44. Which one of the following chemical is now discouraged for seed treatment due to their detrimental effects on seed and environment?
(a) Copper compounds (b) Mercury compounds
(c) Silver compounds (d) Sulpher compounds

45. With the help of which one of the following, separation is done based on seed colour?
(a) Gravity separator (b) Hand picking
(c) Indent cylindrical separator (d) Spiral separator

46. The airscreen cleaner operates on the principles except which one of the following?
(a) Aspiration (b) Cleaning
(c) Grading (d) Scalping

47. For a prescribed period of time, the exposure of imbibed seeds to higher/lower temperatures to overcome mechanical or morphological dormancy is termed as which one of the following?
(a) Priming (b) Scalping
(c) Scarification (d) Stratification

48. With bioactive chemicals, which one of the following is a process of breeding crops for enriching the seeds?
(a) Fortification (b) Halogenation
(c) Priming (d) Stratification

49. Which one of the following is the treatment of seeds with an osmotic solution which allows controlled hydration (e.g., polyethylene glycol)?
(a) Fortification (b) Halogenation
(c) Permeation (d) Priming

50. Which one of the following equipment is not used for pre-cleaning/pre-conditioning operations?
(a) Debearder (b) Groundnut decorticator
(c) Maize sheller (d) Scalper

51. In tobacco cropping, which one of the following term is used to designate the harvesting of ripe tobacco leaves from the stalk as they ripen, beginning at the bottom and progressing upward seed?
(a) Fortification (b) Halogenation
(c) Permeation (d) Priming

52. Which one of the following screen opening is not made on perforated sheet metal sieves?
(a) Oblong (b) Rectangular
(c) Round (d) Triangular

53. Soaking of seeds in organic solvents to improve the germination and vigour of the seed by infusion of bioactive chemicals into the seed without altering seed moisture content is known as
(a) Fortification (b) Halogenation
(c) Permeation (d) Priming

54. Which one of the following is true regarding the priming?
(a) Applied for more uniform and rapid germination of certain vegetable seeds.
(b) The seed embryo develops to the point of germination and then is dried.
(c) Both (a) and (b) (d) None of these

55. What is the role of shoes in an air screen cleaner?
(a) Cleaning of the screen openings
(b) Crushing of clay lumps
(c) Vibrating and shaking of screens
(d) None of these

56. To fix the biochemical events, the hydration of the seed to initiate the pre-germinative metabolism followed by dehydration is termed as which one of the following?
(a) Seed dressing (b) Seed fortification
(c) Seed hardening (d) Seed infusion

57. Which one of the following is not similar with others?
(a) Batch dryer (b) Batch–in–bin dryer
(c) Layer–in–bin dryer (d) Step-in–bin dryer

58. Through a carrier or a high molecular solute, controlled hydration of seeds to a level that starts pre-germinative metabolic activity followed by dehydration to check emergence of the radicle is termed as which one of the following?
(a) Seed dressing (b) Seed fortification
(c) Seed infusion (d) Seed priming

59. Arrange the following seed packaging operations in proper sequence:
(i) Attaching labels & tags on seed bags
(ii) Filling of seed bags to an exact weight
(iii) Sewing of seed bags
(iv) Placing leaflets in the seed bags
(a) i-ii-iii-iv (b) ii-iii-iv-i
(c) ii-iv–i-iii (d) iv-iii-ii-i

60. In salt solutions, soaking of seeds is known as which one of the following?
(a) Bio-priming (b) Halo-priming
(c) Osmo-priming (d) Solid matrix priming

61. Which one of the following refers to the seed coating with biological agents like Rhizobium?
(a) Bio-priming (b) Halo-priming
(c) Osmo-priming (d) Solid matrix priming

62. Heated air drying is usually done at air flow rate of ________ CFM/bu
(a) 10-20 (b) 15-40 (c) 30-50 (d) 40-60

63. Enclosing or encapsulation of the small seed with a small quantity of inert (foreign) material to produce a globular unit of standard size is known as which one of the following?
(a) Seed Fortification (b) Seed hardening
(c) Seed pelleting (d) Seed priming

64. In a typical four screen model air screen cleaner, the third screen performs which function among the following?
(a) Close scalping (b) Fine grading
(c) Grading (d) Scalping

65. With all the possible useful active ingredients viz. Insecticide + Fungicide + Micronutrients + Colouring agents etc. the coating of seed with an adhesive is called as which one of the following?
(a) Designer seed (b) Hard seed
(c) Pelleted seed (d) Prime seed

66. In sesame the seed, with which one of the following is dressed to check mesomechochory?
(a) Bacteriacide (b) Fungicide
(c) Insecticide (d) Nematacide

67. For cleaning of round seeds, top and bottom screens should be which of the following combinations?
(a) Oblong hole and Oblong hole
(b) Oblong hole and Rectangular hole
(c) Round hole and Slotted hole
(d) Slotted hole and Round hole

68. From seeds, the seed feeder that obtains water metabolically and releases it into their immediate environment along with waste heat are called as which one of the following?
(a) Internal seed feeder (b) Primary seed feeder
(c) Secondary seed feeder (d) None of these

69. A very serious insect pest is *Callosobruchus spp.* considered for which one of the following crop?
(a) Cereals (b) Oil seed crops
(c) Pulses (d) Vegetatively propagated crops

12.7 Seed Quality Control

1. The crop __________ and __________ are with lowest seed multiplication ratio and highest seed multiplication ratio, respectively.
   (a) Coconut and Sesame (b) Field pea and Soybean
   (c) Groundnut and Coconut (d) Sesame and Field pea
2. Which one of the following test is not included under routine test?
   (a) Germination test (b) Purity test
   (c) Seed health test (d) Test of seed moisture
3. For drawing seed samples from seed bags which one of the following method to be followed?
   (a) By using bin sampler (b) By using sleeve trier
   (c) By using thief trier (d) Hand sampling
4. For a seed lot stored in bulk, which one of the following method/ device to be followed for drawing of samples?
   (a) By using bin sampler (b) By using sleeve trier
   (c) By using thief trier (d) Hand sampling
5. A seed lot of paddy is having 200 bags each of 25 Kg weight. How many numbers of primary samples to be drawn from this lot for quality analysis?
   (a) 10 (b) 15 (c) 18 (d) 20
6. By sowing of hybrid seeds, which one of the following population will be form?
   (a) Heterogeneous population of heterozygous plants
   (b) Heterogeneous population of homozygous plants
   (c) Homogeneous population of heterozygous plants
   (d) Homogeneous population of homozygous plants
7. Which one of the following method of testing seed borne plant viruses involves the use of electron microscope?
   (a) ELISA (b) Latex flocculation test
   (c) Radio Immune Assays (d) SSEM
8. During seed health testing, abundant development of saprophytic molds and storage fungi is an indication of which one of the following?
   (a) Improper processing (b) Improper storage
   (c) Unfavorable harvesting (d) All of these

9. Presence of insect pests in seeds is generally expressed in terms of which one of the following?
   (a) Number per volume of seed
   (b) Numbers per weight of seed
   (c) Percentage by number
   (d) Percentage by weight
10. For detection of internal (hidden) infestation of insects, the seeds can be made translucent by treating with which one of the following?
   (a) Lactophenol (b) NaOH
   (c) Both (a) and (b) (d) None of these
11. During the verification of the seed source, which one of the following colour of the tag should be if a farmer is producing the certified seed of Wheat variety WH 147.
   (a) Azure Blue (b) Bottle Green
   (c) Golden Yellow (d) White
12. A seed lot of mung bean is weighing 3000 Kg. How many minimum numbers of primary samples to be drawn from this lot for quality analysis?
   (a) 10 (b) 15 (c) 18 (d) 20
13. The minimum weight of working sample for purity analysis is usually ____% of the weight of the submitted sample.
   (a) 5 (b) 10 (c) 15 (d) 20
14. The minimum weight of submitted sample and working sample for purity analysis of maize seed is ______ & ______ g, respectively:
   (a) 1000 & 100 (b) 1000 & 1000
   (c) 1000 & 900 (d) 900 & 900
15. The minimum weight of submitted sample and working sample for purity analysis of sunflower (OPV) seed is ______ & ______ g, respectively:
   (a) 1000 & 100 (b) 1000 & 1000
   (c) 1000 & 250 (d) 1000 & 500
16. The minimum weight of working sample for purity analysis is prescribed as 0.5 g in which one of the crop seed?
   (a) Amaranthus (b) Doob grass
   (c) Tobacco (d) None of these

17. By breeding procedure, name given to all the cultivated variants of plant varieties produced after its notification is designated as which one of the following?
(a) Cultigen (b) Cultivar
(c) EDV (d) Extant variety

18. A minimum of ______ g seed sample is to be submitted to the seed testing laboratory for moisture test for a species that has to be ground:
(a) 50 (b) 100 (c) 150 (d) 200

19. A minimum of ______ g seed sample is to be submitted to the seed testing laboratory for moisture test for a species that has not to be ground:
(a) 50 (b) 100 (c) 150 (d) 200

20. The minimum weight of submitted sample for determination of genuineness of maize cultivars under laboratory test is ______ g.
(a) 500 (b) 1000 (c) 1200 (d) 2000

21. The minimum weight of submitted sample for determination of genuineness of wheat cultivars under both laboratory and field test is ______ g.
(a) 500 (b) 1000 (c) 1200 (d) 2000

22. The low volume weight of seed does not indicate which one of the following?
(a) Dead seed (b) Immaturity
(c) Insect damage (d) Sterility

23. Which one of the following is designating to any genera with only one known species that to under cultivation?
(a) Cultigen (b) Genera
(c) Cultivar (d) Variety

24. Which one of the following vigour test is usually recommended for corn?
(a) Brick-gravel test (b) Cold test
(c) Cool germination test (d) Paper piercing test

25. Which one of the following is the name of the most accurate method of determination of moisture content of seed?
(a) Electrical moisture meter method
(b) Karl-Fisher method
(c) NMR method
(d) Phosphorus pentoxide method

26. Which one of the following is the name of the standard method of determination of moisture content of seed?
(a) Air oven method (b) Electrical moisture meter method
(c) Karl-Fisher method (d) NMR method

27. For sowing purpose to produce Foundation seed, a farmer has to purchase the seed from which one of the following?
(a) Open Market (b) Seed Certification Agency
(c) Seed Corporation (d) University

28. While sampling, if the seed lot does not look homogenous which one of the following test to be conducted to check its homogeneity?
(a) Heterogeneity test (b) Homogeneity test
(c) Physical purity test (d) None of these

29. For sowing purpose to produce certified seed, a farmer has to purchase the seed from which one of the following?
(a) Market
(b) Seed Certification Agency
(c) Seed Corporation
(d) University

30. Which of the following is not considered as multiple seed unit?
(a) Florets with attached single fertile/sterile floret that does not extend to the tip of the fertile floret.
(b) One fertile floret and one fertile/sterile floret attached that extends to or beyond the tip of the fertile floret excluding awn.
(c) One fertile floret with basically attached sterile floret of any length.
(d) One fertile floret with more than one attached fertile/sterile florets of any length.

31. During purity analysis, seeds of Legumes and Crucifers with the seed coat partly removed are classified as which one of the following?
(a) Inert matter (b) Other crop seed
(c) Pure seed (d) Weed seed

32. While conducting purity analysis of a paddy seed sample, a broken seed is found to be more than one half of the original size but without embryo. To which one of the following components it is to be included?
(a) Inert matter (b) Other crop seed
(c) Pure seed (d) Weed seed

33. While conducting purity analysis of a paddy seed sample, a broken seed is found to be less than one half the original sizes but with embryo. It is to be included under which one of the following component?
(a) Inert matter (b) Other crop seed
(c) Pure seed (d) Weed seed

34. The seeds for which light is prescribed for germination should be illuminated for at least which one of the following hours per day during germination test period?
(a) 6 (b) 8 (c) 10 (d) 12

35. Which of the following is not expressed as numbers per kilogram?
(a) Designated inseparable other crop seed
(b) Designated objectionable weed seed
(c) Huskless paddy seeds
(d) Seeds of other distinguishable varieties

36. Determination of genuineness of crop varieties by study of phenol colour reaction is extensively done in which one of the following?
(a) Maize (b) Pearl millet (c) Sorghum (d) Wheat

37. In modified phenol test, the seeds are soaked in dilute solution of which one of the following chemical before putting in phenol solution?
(a) $CuSO_4$ (b) KH_2PO_4 (c) Na_2HPO_4 (d) NaOCl

38. Which one of the following monitor to the Breeder Seed Production programme at field level?
(a) Certified by Field Inspector
(b) Certified by the committee constituted for the purpose
(c) Certified by the Officer of the CSTL
(d) Monitored by the committee constituted for the purpose

39. The formula for calculation of seed vigour index formulated by Abdul-Baki & Anderson (1973) was initially given for which one of the following crop?
(a) Maize (b) Rice
(c) Soybean (d) Sunflower

40. Which one of the following seed health test gives an indication of the viability status of pathogens?
(a) Direct examination of seeds for presence of symptoms
(b) Examination of pathogens removed by washing
(c) Moist blotter method (d) Nome of these

41. Which one of the following method of testing seed borne bacterial pathogens is different from others in respect of mode of action?
(a) Agglutination method
(b) Double diffusion technique
(c) Immuno-fluorescence technique
(d) Phase plaque technique

42. In seed multiplication, how many numbers of tiers is/are involved in India?
(a) One (b) Two (c) Four (d) Three

43. Coarse grinding of seeds for moisture determination is done in which one of the following crops?
(a) Cereals (b) Cotton (c) Legumes (d) Oilseeds

44. Fine grinding of seed s for moisture determination is done in which one of the following crops?
(a) Cotton (b) Legumes (c) Oilseeds (d) Small grains

45. The composite sample is reduced to the required quality with the help of which one of following?
(a) Divider (b) Grader (c) Hand (d) Trier

46. The highest seed replacement rate is of which one of the following
(a) Composite (b) Hybrid (c) Synthetic (d) Variety

47. Pre-drying of seeds before grinding is done if the seed moisture content exceeds:
(a) 10 (b) 13 (c) 15 (d) 17

48. Pre-drying of rice seeds before grinding is done if the seed moisture content exceeds:
(a) 10 (b) 13 (c) 15 (d) 17

49. Pre-drying of soybean seeds before grinding is done if the seed moisture content exceeds:
(a) 10 (b) 13 (c) 15 (d) 17

50. Pre- and post-control tests are made for checking of which quality parameter of seed?
(a) Genetic purity (b) Germination %
(c) Physical purity (d) Seed borne disease %

51. Determination of moisture content of seeds by use of electrical moisture meter is based on which one of the following?
(a) Difference in loss of moisture (b) Electrical conductivity
(c) Seed size (d) Temperature of seed

52. Which one of the following distinguishes to seed and grain?
(a) Genetic purity (b) Seed processing
(c) Seed quality (d) None of these

53. Top of paper (TP) method of germination is usually recommended for which one of the following?
(a) Large size seeds (b) Medium size seeds
(c) Small and Medium size seeds (d) Small size seeds

54. Between paper (BP) method of germination is usually recommended for which one of the following?
(a) Large size seeds (b) Medium and Large size seeds
(c) Medium size seeds (d) Small size seeds

55. Sand (S) method of germination is usually recommended for which one of the following?
(a) Large size seeds (b) Medium and Large size seeds
(c) Medium size seeds (d) Small size seeds

56. In India, the release of new crop varieties consists of four major steps. Which one of the following is a correct arrangement for order of the step?
(1) Development of new strains
(2) Evaluation of performance
(3) Identification of superior strains
(4) Release and notification
(a) 4-3-2-1 (b) 2-3-1-4 (c) 3-1-2-4 (d) 1-2-3-4

57. Which one of the following term is used for obtaining optimum population of healthy plants of the selected variety adopting normal recommended seed rate?
(a) Genuine seed (b) Planting value of seed
(c) Real value of seed (d) Seed aptitude

58. For determination of cultivar purity of soybean, the peroxidise activity of which one of the following part of the seed is tested?
(a) Cotyledon (b) Embryo
(c) Endosperm (d) Seed coat

59. Presence or absence of tannic acid in the seeds of sorghum varieties can be tested by which one of the following test?
(a) Alkaloid test (b) KOH-bleach test
(c) NaOH test (d) Potassium dichromate test

60. Which one of the following cultivar purity test is based on the banding pattern of proteins?
(a) Chromatography (b) Grow-out test
(c) Image analysis (d) PAGE

61. In conducting Grow-out test, a minimum of _____ number of plants need to be observed as which one of the following?
(a) 200 (b) 400 (c) 800 (d) 1000

62. Which one of the following governs to the seed yield from a unit area?
(a) Genetic potential of the variety and environmental condi tion
(b) Genetic potential of the variety, optimum population of healthy plants
(c) Optimum plant population and environmental conditions
(d) Optimum plant population of a potential variety

63. In botanical nomenclature a variety is a sub group of which one of the following?
(a) Class (b) Genera
(c) Kingdom (d) Species

64. Which one of the following is not a performance test of seed vigour?
(a) First count (b) Rate of respiration
(c) Seedling growth rate (d) Speed of germination

65. Which one of the following is not a biochemical test of seed vigour?
(a) Controlled deterioration test (b) GADA test
(c) Leach ate conductivity test (d) Tetrazolium test

66. A plant grouping except micro-organism with in a single botanical *taxon* of the lowest known rank, is called as which one of the following?
(a) Land race (b) Species
(c) Strain (d) Variety

67. Which one of the following is true statements, except?
(a) Breeder seed is used for the production of genetically pure, foundation, registered, and certified seeds.
(b) Certified seed is the progeny of breeder, select, foundation, or registered seed.
(c) Foundation seed is a sort of certified seed, either directly or through registered seed.
(d) Foundation seed represents the true pedigree of the variety.

68. In germinating seed, which one of the following disaccharides sugar found in large amount?
(a) Cellobiase (b) Lactose (c) Maltose (d) Sucrose

69. As per the ISTA rule, the seed germination test should be made on which one among the following?
(a) 200 normal seeds (b) 200 pure seeds
(c) 400 normal seeds (d) 400 pure seeds

70. Which one of the following is not a seed counting device?
(a) Electronic seed counter (b) Hand seed counter
(c) Seed counting board (d) Vacuun seed counter

71. The particle size of the sand used for seed germination should be within the range of which one among the following?
(a) 0.05 and 0.8 mm (b) 0.08 and 0.5mm
(c) 0.5 and 0.8 mm (d) 05 and 0.5 mm

72. Based on the result of the laboratory established by the producer, the seed sold but not from a recognized lab is considered as which one of the following?
(a) Certified seed (b) Denotified seed
(c) Privately Certified seed (d) Truthfully labeled seed

73. The lactophenol solution used as a clearing agent in TZ test of seed viability is prepared by mixing lactic acid, phenol, glycerine and water, respectively in which one of the following proportion?
(a) 1:1:1:1 (b) 1:1:1:2 (c) 1:1:2:1 (d) 1:2:1:1

74. To facilitate the entry of tetrazolium solution, the seed embryo is bisected laterally in which one which one of the following seeds?
(a) Corn (b) Large seeded grasses
(c) Small grains (d) Small seeded grasses

75. In a TZ test of seed viability in rice, it was observed that the extremities of scutellum, radicle tip and coleorhizae are unstained. The seed will be evaluated as which one of the following?
(a) Dead (b) Dormant
(c) Germinable (d) Non-germinable

76. In the TZ test of seed viability of a non-legume dicot seed, small necroses at the extreme tip of radicle was observed. The seed will be evaluated as which one of the following?
(a) Dead (b) Dormant
(c) Germinable (d) Non-germinable

77. The embryo excision method is generally recommended for test of seed viability in which one of the following categories of seeds?
(a) Cucurbits (b) Ornamentals
(c) Spices (d) Tree species

78. In India, which one of the following class that does not represent to the seed multiplication chain?
(a) Nucleus seed (b) Breeder seed
(c) Certified seed (d) Foundation seed

79. The generation between Foundation and Certified seed mainly for autogamous crops in United States is considered as which one of the following?
(a) Certified seed I (b) Foundation seed II
(c) Registered seed (d) Truthfully labeled seed

80. In which one of the following crops, pre-chilling of seeds is not effective in breaking dormancy?
(a) Jute (b) Lentil (c) Mustard (d) Sorghum

81. In which one of the following grass seeds, exposure of seeds to light is not essential to effect germination?
(a) Carpet grass (b) Dallis grass
(c) Doob grass (d) Para grass

82. Which one of the following organization plays the role for production and procurement of Breeder, Foundation and Certified seed?
(a) Agricultural University (b) Seed Certification Agency
(c) Seed Corporation (d) None of these

83. In seed multiplication chain, how many numbers of autonomous bodies are involved?
(a) Two (b) Four (c) Three (d) Five

84. The seeds for which light is prescribed for germination should be provided at an intensity within the range of which one among the following?
(a) 600 – 1000 Lux (b) 700 – 1200 Lux
(c) 750 – 1200 Lux (d) 750 – 1250 Lux

85. Which one of the following method of germination should be followed for light requiring seeds?
(a) BP method (b) PP method
(c) S method (d) TP method

86. Pre-chilling of seeds for breaking dormancy should be done at a temperature of ___ ^{0}C for ___ days.

(a) 0 – 10 ^{0}C for 10 days (b) 10 ^{0}C for 10 days

(c) 5 – 10 ^{0}C for 7 days (d) 5 ^{0}C for 10 days

87. In pulses, the germination percentage is calculated on the basis of which one of the following?

(a) Normal seedlings + Abnormal seedlings

(b) normal seedlings + Abnormal seedlings + Hard seeds

(c) Normal seedlings + Hard seeds

(d) Normal seedlings only

88. From Breeder - Foundation – Certified, how many number of seed cycle are multiplied and then to food purpose by the farmer?

(a) Two (b) Four (c) Three (d) None of these

89. Which one of the following is not a category of normal seedling?

(a) Intact seedling

(b) Seedling with primary infection

(c) Seedling with slight defect

(d) None of these

90. Which one of the following is not a category of abnormal seedling?

(a) Damaged seedling (b) Deformed seedling

(c) Seedling with secondary infection (d) Unbalanced seedling

91. How long period, seed certificate is valid?

(a) 12 month from the date of sowing of the seed crop

(b) Nine months from the date of harvest

(c) Nine months from the date of processing

(d) Nine months from the date of test

92. The first and final count day of taking observation in the germination test of rice seed are:

(a) 10 & 15 (b) 5 & 10

(c) 5 & 14 (d) 7 & 14

93. The first and final count day of taking observation in the germination test of pigeon pea seed are:

(a) 10 & 15 (b) 4 & 6 (c) 5 & 10 (d) 7 & 14

94. The first and final count day of taking observation in the germination test of rape and mustard seed are:

(a) 10 & 15 (b) 5 & 10 (c) 5 & 7 (d) 7 & 14

95. The first and final count day of taking observation in the germination test of cotton seed are:
(a) 10 & 15 (b) 4 &12 (c) 5 & 10 (d) 5 & 14

96. Under the supervision of Seed Certification Agency, a registered seed growers produces the seed, which is certified by a blue colour (Shade ISI No. 104, azure blue). This seeds are categorized under which one of the following seed category?
(a) Breeder seed (b) Certified seed
(c) Foundation seed (d) Registered seed

97. Under which one of the following conditions the germination test shall be repeated?
(a) Error in evaluation and calculation
(b) Persistence of dormancy
(c) Replicate variation is out of tolerance
(d) All of these

98. According to which one of the following scientists, the tetrazolium test of seed viability can be conducted with a representative sample of 50 or 100 seeds?
(a) Delouche et. al (1962) (b) Grabe (1970)
(c) Justice (1972) (d) Moore (1960)

99. Who one of the following gave the first clear-cut concept and definition of seed vigour?
(a) Frank (1950) (b) Hiltner (1917)
(c) Isley (1957) (d) Ista (1972)

100. Who one among the following scientist had developed the Brick-gravel test of seed vigour?
(a) Frank (1950) (b) Hiltner (1917)
(c) Isley (1957) (d) Ista (1972)

101. The seed crop as per seed act, should be raised from the seed of which one of the following?
(a) Approved source
(b) Genetic pure seed produced by the seed grower
(c) Genetic pure seed purchased from the open market
(d) Genetic pure seed without any approved source

102. Which one of the following seed vigour test is used for prediction of relative storability of seed?
(a) Accelerated ageing test (b) Brick-gravel test
(c) Leachate conductivity test (d) Paper piercing test

103. Which one of the following test is used for assessment of both seed viability and vigour?
(a) Controlled deterioration test (b) GADA test
(c) Membrane integrity test (d) TZ test

12.8 Seed Storage

1. Without damage, which one of the following seeds may withstand dehydration?
(a) Dormant seed (b) Hard seed
(c) Orthodox seed (d) Recalcitrant seed

2. Which one of the following factors is considered the most important and crucial in affecting seed quality in storage?
(a) Moisture content of seeds (b) Relative humidity
(c) Storage pests (d) Storage temperature

3. Which one of the following environmental factors are the two greatest enemies of storage life of seeds?
(a) High light intensity and High temperature
(b) High moisture and High temperature
(c) Low light intensity and High moisture
(d) Low oxygen and High moisture

4. The equilibrium moisture content of seed at a given RH tends to increase as which one of the following?
(a) At constant temperature (b) Temperature decreases
(c) Temperature has no effect (d) Temperature increases

5. The maximum size based on weight of true seed in a seed lot is of which one of the following field crops?
(a) Maize (b) Rice (c) Soybean (d) Wheat

6. At a constant temperature, if the seed moisture content is decreased by one percent then the longevity of seed will which one of the following?
(a) Doubled (b) Quadrupled (c) Remain same (d) Tripled

7. Cryopreservation refers to the storage (conserving) of materials at which one of the following?
(a) Liquid nitrogen temperature (b) Low temperature
(c) Room temperature (d) Very high temperature

8. Good storage is achieved at which one of the following conditions?
(a) Relative humidity(%) + temperature (°F) < 100
(b) Relative humidity(%) + temperature (°F) add up to 100
(c) Seed moisture content(%) + temperature (°F) < 100
(d) Seed moisture content(%) + temperature (°F) add up to 100

9. Storage or conservation of materials at –196 °C is known as which one of the following?
(a) Cold storage (b) Cryopreservation
(c) Medium term storage (d) Short term storage

10. Storage fungi are completely inactive below which one of the following relative humidity (RH)H?
(a) 52 % (b) 62 % (c) 72 % (d) 75 %

11. Optimal storage condition refers as which one of the following?
(a) Airtight, High humidity, and Low temperature
(b) Airtight, Low humidity, and High temperature
(c) Airtight, Low humidity, and Low temperature
(d) Free air flow, Low humidity, and Low temperature

12. The insect-pests activity in seed storage ceases at seed moisture content below which one of the following?
(a) 6 (b) 8 (c) 10 (d) 12

13. An improvement in seed performance by any post-harvest physio-chemical treatment resulting in better storability, improved germination and field performance over a wide range of edapho-climatic conditions than the corresponding untreated seed is called as which one of the following?
(a) Seed ageing (b) Seed invigoration
(c) Seed pelleting (d) Seed pelleting

14. The storage temperature at a range of which one of the following ? are considered as unsafe for insect activity?
(a) 10 – 15 (b) 15 – 20 (c) 17 – 22 (d) 5 – 10

15. For effective seed drying and maintenance of seed moisture content, seed should be packed in moisture-proof container along with desiccant at a proportion of which one of the following
(a) 5:1 (b) 10:1 (c) 12:1 (d) 15:1

16. The optimum height of stacking seed bags in a store house is which one of the following?
(a) 15-20 Ft (b) 20-25 Ft
(c) 5 -10 Ft (d) 10-15 Ft

17. Elements of Group XVII of periodic table are known as Halogen. Halogenation is the replacement of one atom of Hydrogen. The halogen is absorbed by of the seed and reduces the physiological deterioration of which one of the following?
(a) Carbohydrate (b) Protein
(c) Saturated fatty acid (d) Unsaturated fatty acid

18. Providing a 15 cm lip around the building at 90 cm level of the floor is effective in controlling which one of the following storage pest?
(a) Birds (b) Insects (c) Rodents (d) All of the above

19. During storage, the halogen protect seed from the action of which one of the following?
(a) Antimicrobial (b) Attractan
(c) Fumigation (d) Repellent

20. Which one of the following situation is considered unsafe for seed storage?
(a) Cool and dry condition (b) Cool and humid condition
(c) Warm and dry condition (d) Warm and humid condition

21. Prescribed maximum limit of a soybean seed lot is which one of following>
(a) 10,000 kg (b) 20,000 kg (c) 40,000 kg (d) None of these

22. Which one of the following situation is considered ideal for seed storage?
(a) Cool and humid condition (b) Cool and dry condition
(c) Warm and humid condition (d) Warm and dry condition

23. All the stored insect pests belong to which one of the following Order?
(a) Coleopetra, Diptera, and Odonata
(b) Coleopetra, Lepidoptera, Hymenoptera
(c) Hymenoptera, Diptera, and Odonata
(d) Lepidoptera, Diptera and Odonata

24. The carry-over seeds can be safely stored for 12 to 18 months by maintaining proper RH and temperature. Which one of the following option would be effective?
(a) 25% RH & 30 °C temp. (b) 20% RH & 40 °C temp.
(c) 45% RH & 25 °C temp. (d) 455 RH & 30 °C temp.

25. Important store grain pests are from which one of the following class?
(a) Chordeta (b) Insecta and Acarira
(c) Only Acarira (d) Reptilia

26. Provision of which one of the following in the store house can effectively control both temperature and moisture?
(a) Dehumidification (b) Insulation
(c) Refrigeration (d) Ventilation

27. Cigarette beetle, confused flour beetle, Indian meal, red flour beetle and saw toothed beetle are known as
(a) External and internal feeder (b) External feeder
(c) Internal feeder (d) None of these

28. Which one of the following seed quality is not affected during storage?
(a) Genetic (b) Health
(c) Physiological (d) None of these

29. Seeds already damaged by insects in store are fed upon by which one of the following?
(a) Mites (b) Nematodes
(c) Rat (d) Snail

30. Which one of the following fumigant needs a relatively longer exposure period for effective control of insect pests in storage?
(a) Hydrogen cyanide (b) Hydrogen phosphide
(c) Methyl bromide (d) None of these

31. To control seed borne fungi and insects, to control deterioration of seed during storage et. Which one of the following term is used for the exposure of seeds to gaseous form of harmful chemicals?
(a) Fumigation (b) Halogenation
(c) Invigoration (d) Pelleting

32. Out of the following sources of heat in a seed store house, which has the least effect?
(a) Field heat (b) Heat leakage
(c) Heat of respiration (d) Incidental heat

33. In case of which one of the following, more than the 1% insect infested seeds are rejected?
(a) Cereals (b) Legume and Maize
(c) Oil seed (d) Vegetable

34. Silica gel and activated alumina are used as solid desiccants in seed storage because of which one of the following properties?
(a) Cheap and Efficient (b) Indestructible
(c) Non-toxic (d) All of the above

35. Maximum permitted insect infestation during storage is which one of the following in oil seed crops?
(a) 0.1% (b) 0.5% (c) 1% (d) None of these

36. Silica gel can absorb moisture up to which one of the following % of its dry weight?
(a) 30 (b) 40 (c) 50 (d) 60

37. Primary sample of rice stored in bags is drawn with the help of which one of following?
(a) Seed Divider (b) Trier
(c) Hand (d) Cup

38. Used silica gel can be reactivated by drying it in an oven at a temperature of which one of the following ??
(a) 100 (b) 150
(c) 175 (d) 200

39. Which one of the following is an objectionable insect pest of sweet potato?
(a) Bruchus (b) Scale and Mealy bug
(c) Scale insect (d) Wireworm and Weevil

40. Which one of the following chemical can be used as a liquid spray desiccant?
(a) Lithium bromide (b) Lithium chloride
(c) Lithium oxide (d) All of the above

41. Which one of the following crops considered to the scale insect as an objectionable pest?
(a) Colocassia (b) Okra
(c) Potato (d) Sugarcane

42. The granules of silica gel are treated with which one of the following chemicals to make itself indicating?
(a) Cobalt chloride (b) Cobalt nitrate
(c) Lithium Chloride (d) Lithium oxide

43. Seed of wheat reduced to husk indicates to which one of the following infestation?
(a) Flat grain beetle (b) Khapra beetle
(c) Red flour and Beetle (d) Saw toothed beetle

12.9 Seed Certification

1. In 1919 the International Crop Improvement Association (ICI(A) was formed and renamed as Association of Seed Certification Agencies (AOSC(A) in which one of the following year?
(a) 1966 (b) 1967 (c) 1968 (d) 1969

2. Which of the following number indicates the phases involved in the seed certification process?
(a) 5 (b) 6 (c) 7 (d) 8

3. The husk less seed of the following crops are counted separately in physical purity analysis?
(a) Sorghum and Pearl millet (b) Oat and Barley
(c) Sunflower and Rice (d) Niger and Sesame

4. Whenever more than two field inspections are prescribed for a crop, the first inspection is invariably done before:
(a) Flowering (b) Harvesting (c) Maturity (d) Sowing

5. Foundation seed producer transfers the seed to which one of following bodies?
(a) Seed Certification Agency (b) Seed Corporation
(c) Seed Processing Plant (d) None of these

6. Indicate which of the following is not satisfying the basic principle of forming a Seed Certification Agency?
(a) It should adopt uniform procedure throughout the country.
(b) It should be directly under the control of State Department of Agriculture.
(c) It should operate with no profit-no loss basis.
(d) The SCA should not be involved in production and marketing of seed.

7. Minimum rotation gap required for seed production programme of different varieties of the same crop in the selected field is
(a) 1-2 year (b) 2-3 years (c) 3-4 (d) 4-5 year

8. Bin inspection is related to seed production in which of the following crops?
(a) Cotton (b) Maize (c) Rice (d) Sunflower

9. If in the Last year/season, the field will not be selected for seed production programme because of__________.
(a) Different crop was cultivated
(b) Different variety of the same crop was cultivated
(c) No crop was cultivated
(d) Same variety of the same crop was cultivated

10. Inter cropping during seed certification is permitted in which of the following group of crops?
(a) Cereals and pulses (b) Cereals and oil seeds
(c) Pulses and oil seeds (d) All of the above

11. The validity period of seed certification could be further extended provided on re-testing seed conforms to the prescribed standards for which one of following?
(a) 3 Months (b) 6 Months
(c) 9 Months (d) 12 Months

12. Under the specific crop standards, the maximum permissible limits of the following have been specified to ensure genetic purity, except:
(a) Off types (b) Pollen shedders
(c) Shedding tassels (d) Volunteer plants

13. Under which one of following, the seed production plot should be?
(a) Inter cropping (b) Mixed cropping
(c) Sole cropping (d) None of these

14. The certificate issued by the SCA remains valid for how many months?
(a) 6 (b) 9 (c) 12 (d) 15

15. Which one of following term state, if insect preset in the seed sample?
(a) Inert matter (b) Bio matter
(c) Pure seed (d) Not a part of seed sample

16. Random method of field inspection is recommended for which seed crop?
(a) Maize (b) Pearl millet
(c) Rice (d) Sorghum

17. Which one of following term is used for plants formed by the seed of the crop grown last year in the same field?
(a) Objectionable weed (b) Off type
(c) Rogue (d) Volunteer plant

18. The clock-wise pattern of field inspection is recommended for which group of crops?
(a) Cereals (b) Oilseeds
(c) Pulses (d) None of these

19. The validity period of a seed testing certificates is for how long?
(a) 12 month (b) 15 month (c) 6 month (d) 9 month

20. The validity period of certificate issued by the SCA can be further extended for ______ months provided the seed s confirms to the prescribed quality standards.
(a) 6 (b) 9 (c) 12 (d) 15

21. The validity period of seed certification could be further extended provided on re-testing seed conforms to the prescribed standards in respect of which one of following?
(a) Genetic purity, physical purity and germination
(b) Moisture content, germination and seed health
(c) Physical purity, germination and insect damage
(d) Validity period cannot be extended

22. How many counts need to be taken from a seed plot of 4.5 ha?
(a) 5 (b) 6 (c) 7 (d) 8

23. The colour of the certification tag issued to the Foundation seed is:
(a) Blue (b) Green (c) White (d) Yellow

24. The colour of the certification tag issued to the Certified seed is:
(a) Blue (b) Green (c) White (d) Yellow

25. The lowest physical purity percentage for seed certification is required among which one of the following crops?
(a) Groundnut (b) Wheat (c) Soybean (d) Pearl millet

26. It is mandatory to conduct the Grow-out Test for the hybrid seeds produced by which of the following methods?
(a) Detasseling (b) Use of CGMS
(c) Use of CHAs (d) Use of GMS

27. From seed production plot, when the off-type and rogues should be removed?
(a) After flowering (b) At the time of maturity
(c) Before flowering (d) Before sowing

28. The seed crops in which minimum two field inspections are prescribed, the inspections shall be made at:
(a) Flowering to harvesting (b) Maturity to harvesting
(c) Pre-flowering to maturity (d) Seeding to maturity

29. GOI in proforma of which one of following, the report of the monitoring team is submitted to ICAR and Seed Division Ministry of Agriculture?
(a) BSP I (b) BSP II (c) BSP III (d) BSP V

30. Which one of following may deteriorated by seed production plot, isolation distance, volunteer plants, presence of off-types and rogue?
(a) Genetic purity (b) Genetical purity
(c) Physical purity (d) None of these

31. The SCA has been empowered to revoke a certificate already issued to a seed producer under Section ______ of the Seed Act (1966).
(a) 7 (b) 8 (c) 9 (d) 10

32. From the seed production plot of *Lens culinaris,* removal of off-type lentil plant refers to which one of following?
(a) Cleaning (b) Nicking
(c) Rouging (d) Weeding

33. The sum total of the part present in the seed sample other than the seed of the crop under test is known as which one of the following?
(a) Inert matter (b) Other part
(c) Dockage (d) Stone

34. Under the system of seed certification, down grading of seed class is permitted in which of the following?
(a) Hybrid varieties (b) Open pollinated varieties
(c) Parental lines (d) None of these

35. Modification in distance to keep the seed crop in isolation is permitted only in hybrid seed production of which one of following crops?
(a) Maize (b) Pearl millet
(c) Pigeon pea (d) Sunflower

36. Under the provision of Section ______ of the Seed Act (1966), the Seed Certification Agencies are established to carry out certification work.
(a) 6 (b) 7 (c) 8 (d) 9

37. Which one of following the tolerance limits for off-types to establish the uniformity Under DUS test for Self-pollinated crops, except cereals crops?
(a) 0.1% (b) 1% (c) 5% (d) 10%

38. The certified Foundation Seed shall be the progeny of:
(a) Breeder Seeds (b) Certified Seeds
(c) Foundation Seeds (d) Both (a) and (c)

39. The head quarter of PPV and FR is located which one of the following place in India?
(a) Bangloere (b) Hyderabad
(c) Lucknow (d) New Delhi

40. The seed lots of maize and legume under certification shall not have visible evidence of insect damage in excess of ______
(a) 0.5% (b) 1.0% (c) 1.5% (d) 2.0%

41. The presence of red flower on the same cotton plant with white flower is considered as which one of following?
(a) Genetic pure (b) Off-type
(c) Out crossed (d) Rogue

42. Partial rejection of seed field is done due to which one of the following?
(a) Lodging of plants.
(b) Maintenance of improper isolation.
(c) Presence of contaminants beyond the maximum permissible limit.
(d) None of these

43. Complete rejection of seed field is done due to which one of the following?
(a) Lodging of plants. (b) Maintenance of improper isolation.
(c) Presence of contaminants beyond the maximum permissible limit.
(d) None of these

44. Scattered fields constituting one unit of seed certification should not be separated by which one of following?
(a) More than 100 m (b) More than 40 m
(c) More than 50 m (d) More than 500 m

45. For seed certification, number of field inspection rages from which one of following?
(a) 1-2 (b) 1-5 (c) 2-4 (d) 5-10

46. After receiving a sample, a seed testing laboratory shall deliver the test report to the sender of the sample not later than ______ days.
(a) 21 (b) 30
(c) 35 (d) After completion of all the tests

47. For India, seed certification standard is fixed by which one of the following agencies?
(a) AOSSCA (b) Central Seed Certification Board
(c) ISTA (d) PPV and FRA

48. Which one of following is type of deterioration indicate due to loss in genetic purity?
(a) Crop deterioration (b) Seed deterioration
(b) Soil deterioration (d) Variety deterioration

49. Under the supervision of the officer from which one of following bodies, the Certified seed is processed?
(a) Engineer of Seed Processing Plant
(b) Seed Certification Agency
(c) Seed Corporation
(d) ICAR institute/ ICAR University

50. Deterioration of varieties due to mutation such as 'fatuoids' in oats or 'rabbit ear' in peas can be controlled by
(a) Change in the seed production field
(b) Production of seed in isolation
(c) Rouging (d) Seed treatment

51. A farmer shows deterioration of which one of following, dur to presence of objectionable weed in the seed of rice produced?
(a) Crop (b) Seed (c) Soil (d) Variety

52. The highest physical purity percentage for seed certification is required among which one of the following crops?
(a) Wheat (b) Rice (c) Cabbage (d) Okra

53. The certified Seeds shall be the progeny of:
(a) Breeder Seeds (b) Certified Seeds
(c) Foundation Seeds (d) Both (b) and (c)

54. Which one of following term is used for number of plants/ear heads of seed crop which should be observed during field inspection as one unit?
(a) Field Reckon (b) Field Count
(c) Field Step (d) Field Assessment

55. Rouging and cultivation of crop in isolation are the effective instruments to check deterioration of which one of following purity?
(a) Physical (b) Genetic
(c) Crop (d) Plant

56. Which one of following is designated inseparable crop plants during seed production of wheat?
(a) Chickpea (b) Lentil
(c) Mung bean (d) Oat

57. Which of the following numbers indicate the parts in a seed lot number?
(a) 2 (b) 3 (c) 4 (d) 5

58. A breeder applied the method for development of a variety may influence deterioration of which one of following?
(a) Crop (b) Seed
(c) Seed health (d) Variety

59. Where is ISTA located?
(a) Munich (b) New Delhi
(c) Rome (d) Zurich

60. The seed varieties eligible for certification shall be which one of following?
(a) Identified by the respective workshop of the crop
(b) Notified under section 5 of the seeds Act, 1966
(c) Protected under PPV and FR Act, 2001
(d) Released by CVRC

61. The maximum seed lot size for maize is _______ Kg.
(a) 10,000 (b) 20,000 (c) 30,000 (d) 40,000

62. In the field, the plant of the same crop present due to previous year/season crop is called as which one of following?
(a) ODV (b) Shattered plant
(c) Volunteer plant (d) Weed

63. While taking observation on the extent of contaminants from a maize seed field, how many plants to be observed per single count?
(a) 100 (b) 500 (c) 1000 (d) None of these

64. While taking observation on the extent of contaminants from a cotton seed field, how many plants to be observed per single count?
(a) 100 (b) 500 (c) 1000 (d) None of these

65. While taking observation on the extent of contaminants from a mustard seed field, how many plants to be observed per single count?
(a) 100 (b) 500 (c) 1000 (d) None of these

66. While taking observation on the extent of contaminants from a soybean seed field, how many plants to be observed per single count?
(a) 100 (b) 500 (c) 1000 (d) None of these

67. With different expression mainly for phenological traits, plant of same variety is removed from the seed production plot to reduce which one of following?
(a) Genetic drift (b) Genetic erosion
(c) Genetic identity (d) Genetic shift

68. The breeder seeds are issued with _______ colour labels.
(a) Blue (b) Green (c) White (d) Yellow

69. Where is Central seed testing laboratory located?
(a) Banglore (b) Hyderabad
(c) New Delhi (d) Varanasi

70. The maximum field area that can be taken under one unit of certification is:
(a) 2 ha (b) 5 ha (c) 10 ha (d) 12 ha

71. Between two which one of following, isolation distance is maintained?
(a) Crops
(b) Cross incompatible species of the same genera
(c) Genera (d) Varieties of the same crop

72. At what one of following level, certification of seed is done?
(a) One level (b) Two levels
(c) Three levels (d) Four levels

73. During field inspection, percent field refers to which one of following percent should be covered?
(a) >90% (b) About 25-40%
(c) About 50% (d) About 60-80 %

74. Bulking of unprocessed seed stocks to obtain larger homogenous stock is permitted by the SCA under the following conditions except:
(a) Belong to the same producer.
(b) Have more or less similar physical appearance.
(c) Of same crop, variety, and class of seed.
(d) Produced in any season but under similar agro-climatic condition.

75. On the basis of field size, number of field counts required ranges from which one of following?
(a) 2-7 (b) 5-9 (c) 7-10 (d) 10-20

12.10 Seed Law Enforcement

1. In the year 1961, which one of the following professor visited India and emphasized on controlling the quality of seeds by certifying them and enacting seed law?
(a) A.S. Carter (b) K. Dorph Peterson
(c) L.O. Copeland (d) O.L. Justice

2. In which one of the following the year, minimum Seed certification were determined?
(a) 1966 (b) 1971 (c) 1975 (d) 1989

3. On matter arising out of the administration of the Seed Act, which one of the following committee advised the Central Government and the State Governments?
(a) Central Seed Committee (b) ICAR Committee
(c) State Seed Committee (d) University Committee

4. For seed testing, the method to be adopted by which one of the following is finalized?
(a) Central Seed Testing Lab
(b) Directorate of Seed Research
(c) International Seed Testing Association (IST(A)
(d) National Seed Testing Association

5. In which one of the following year, the seed rules were passed?
(a) 1963 (b) 1966 (c) 1968 (d) 1971

6. Seed inspector follow which one of the following procedure for inspection of seed, is described in the book entitled?
(a) A hand book of seed inspector
(b) Seed Certification manual
(c) Seed Technology
(d) Seed Testing manual

7. In Seed Testing laboratory, one who of the following agency takes the sample of the notified variety being sent for testing?
(a) Seed inspector of seed certification agency
(b) Officer of seed corporation
(c) Engineer of processing plant
(d) Representative of the University

8. Which one of the following seed was included in the commodities by the Seed (control) order?
(a) Essential (b) Food
(c) Non-essential (d) Volatile

9. In public sector quality of seed is the responsibility of one of the following?
(a) Farmer (b) Seed Certification Agency
(c) Seed Corporation (d) Seed seller

10. In which one of the following year, the National Seed Policy (NSP) was framed?
(a) 1966 (b) 1983 (c) 2002 (d) 2006

11. One or more related species or sub-species of crop plants each individually or collectively known by one common name in seed act 1966 is termed as which one of the following?
(a) Brand (b) Kind (c) Type (d) Variety

12. The seed testing laboratory to which the sample has been submitted for analysis shall submit the report of analysis to the Seed Inspector within how long days of receipt of the sample?
(a) 30 days (b) 45 days (c) 60 days (d) 90 days

13. The person, as per Seed Act 1966 who contravenes the functioning of Seed Inspector from exercising powers for the first offense.may fine upto rupees?
(a) Rs 500 (b) Rs 1000 (c) Rs. 5000 (d) Rs. 10000

14. The sowing will be done with which one of the following class of seed for the production of foundation seed?
(a) Breeder seed (b) Certified seed
(c) Foundation seed (d) Nucleus seed

15. Production of the Breeder seed is the responsibility of
(a) Central Seed Testing Board
(b) Division of Seed, Ministry of Agriculture, Government of India
(c) National Seed Corporation
(d) State Government

16. Which one of the following bodies is responsible for availability of foundation seed for production of certified seed?
(a) Central Seed Testing Board
(b) Division of Seed, Ministry of Agriculture, Government of India
(c) ICAR (d) State Government

17. A national seeds policy has been formulated in which one of the following years to facilitate development, production and distribution of improved varieties of seeds and planting materials?
(a) 2000 (b) 2001 (c) 2002 (d) 2003

18. Which one of the following is the number of generation allowed after Breeder seed in seed multiplication chain?
(a) 2 (b) 3 (c) 4 (d) 5

19. Which one of the following bodies is responsible for production of Foundation and Certified seed?
(a) ICAR (b) Seed Certification Agency
(c) Seed Corporation (d) Seed Testing Laboratory

20. On the tag of the Breeder Seed Signature is of which one among the following?
(a) Plant Breeder (b) Seed Analyst
(c) Seed certification inspector (d) Seed Certification Officer

21. For Certified seed production, a farmer has need to get registered with which one of the following agency?
(a) Seed Certification Agency (b) Seed Corporation
(c) State Department of Agriculture (d) University

22. The first transgenic hybrid of which one of the following crop was allowed in March 2002, for commercial cultivation in farmer's field in India?
(a) Cotton (b) Maize (c) Pearl millet (d) Tomato

23. Which one of the following type seed is generally used by farmers?
(a) Breeder seed (b) Certified seed
(c) Foundation seed (d) Nucleus seed

24. In which one of the following year, India became member of ISTA?
(a) 1961 (b) 1963 (c) 1966 (d) 1971

25. Under the Essential Commodities Act-1955, the Seeds Control Order was enacted in which one of the following year?
(a) 1981 (b) 1983 (c) 1987 (d) 1988

26. In which one of the following year, new policy on Seed Development was enacted?
(a) 1985 (b) 1988 (c) 1998 (d) 2008

27. In which one of the following year, Industrial Seed Policy was enacted?
(a) 1961 (b) 1971 (c) 1981 (d) 1991

28. Environment Protection Act-1986, pertaining rules to Genetically Modified Organisms (GMOs) was enacted in the year which one of the following?
(a) 1986 (b) 1989 (c) 2002 (d) 2006

29. In which one of the following year, the Regulation of Import into India for Plants, Fruits and Seeds Order were was enacted?
(a) 1987 (b) 1989 (c) 1999 (d) 2005

30. In which one of the following year, Protection of Plant Varieties and Farmer's Rights Act was enacted?
(a) 1999 (b) 2000 (c) 2001 (d) 2005

31. In which one of the following year, Environment Protection Act was enacted?
(a) 1987 (b) 1989 (c) 1999 (d) 2005

32. What was the enacted in the year from the following for National Seed Policy?
(a) 1966 (b) 1968 (c) 1998 (d) 2002

33. To regulate the seed industry, which one of following was enacted in 1966 by the Government of India?
(a) Seed act (b) Seed bill
(c) Seed law (d) Seed regulation

34. Through which one of following independent State seed Agency, the act provided a system for seed quality control?
(a) Department of Agriculture (b) Seed certification
(c) Seed corporation (d) Seed production

35. As which one of following was identified under Industrial Policy seed production in 1991?
(a) Cottage industry (b) High priority industry
(c) Low priority industry (d) Not as an industry

36. Minimum rotation gap required for seed production programme of different varieties of the same crop in the selected field is
(a) One season (b) Five years
(c) One to two years. (d) Two-three years

37. Which one of following act is that provide protection to a variety in India?
(a) IPR on Plant Variety Act
(b) Plant Variety Patent Act
(c) Plant Variety Protection and Farmers Right Act
(d) Protection of Plant Variety and Farmers' Right Act (PPVFR)

38. Which one of following is a term defined by Seed act (1966, sub section 16 of section 2) as a sub division of a kind identified by its growth, yield, plant, fruit, seed or other characters?
(a) Cultivar (b) Farmer variety
(c) Land race (d) Variety

39. A notified under section 5 of Seed Act 1966; and Farmers' variety as defined in PPV act; and a variety about which there is common knowledge or any other variety, which is in public domain under PPVFR act. is known as which one of following?
(a) Extant variety (b) Extend variety
(c) Extent variety (d) Extinct variety

40. Which one of following is a variety in respect to the initial variety when it is predominantly derived from such initial variety?
(a) Essentially derived variety (EDV) (b) Extant variety
(c) Extent variety (d) Old variety

41. Under section-5 for how long years among the following, the variety is notified?
(a) 05 (b) 10 (c) 15 (d) 20

42. Farmers traditionally cultivate a variety and evolved by the' in their fields. It may be a wild relative or land race of a variety about which farmer possess the common knowledge is known as
(a) Extant variety (b) Farmers' variety
(c) Land race (d) Wild variety

43. A variety is notified by the office of which one of following?
(a) ICAR Organization
(b) Ministry of Agriculture Government of India
(c) Ministry of Agriculture State Government
(d) None of these

44. Seed of only which one of following varieties is produced by seed multiplication chain?
(a) Denotified (b) Identified
(c) Notified (d) Released

45. India has adapted how many tier systems from the following for seed multiplication?
(a) One (b) Three (c) Two (d) Four

46. The short form of International Union for the Protection of New Varieties of Plants is which one of following?
(a) ISTA (b) IUPNVP
(c) PPV&FRA (d) UPOV

47. Which one of following is the right for researchers as per PPVFR Act?
(a) Free access to registered varieties for export.
(b) Free access to registered varieties for market.
(c) Free access to registered varieties for research.
(d) No free access to registered varieties for research.

48. Full control on a protected variety is of which one of following?
(a) Authority (b) Breeder
(c) Farmer (d) ICAR

49. Who one of following, has rights of commercialization for the registered variety on grant of protection?
(a) Authority (b) Breeder (c) Farmer (d) ICAR

50. Which one of following for making EDVs required in the Indian Act?
(a) Authorization of PPV and FR Authority is needed.
(b) Only original breeder can develop the EDV.
(c) The breeder's authorization is needed.
(d) The breeder's authorization is not needed

51. Indian act granted by which one of following on the plant variety?
(a) Copyright (b) Exclusive right
(c) Patent (d) Protection

52. As per Indian Act methods and processes of agriculture and horticulture
(a) Can be granted copyright.
(b) Can be granted exclusive right.
(c) Can be patented. (d) Cannot be patented.

53. The Indian Patent Amendment Act, call, cell lines, cell organelles like mitochondria and genes Is there any Act for protecting a new plant variety in India?
(a) The Plant variety Protection Act 2001.
(b) The Protection of Plant Varieties Act 2001.
(c) The Protection of Plant Varieties and Farmers' Rights Act 2001 (PPVFR).
(d) All of the above

54. Under the TRIPS agreement it is obligatory on part of a Member to provide protection to new plant variety therefore India opted for which one of following?
(a) Copy right (b) Exclusive right
(c) Patent (d) sui generis system

55. The safeguards are provided to which one of following by the PPVFR act 2001 of India?
(a) Only to breeders (b) Only to farmers
(c) Only to researchers (d) All of the above

56. Which one of following is the kind of varieties that registered under the PPV & FR Act?
(a) Only extant varieties without confirmation of DUS testing
(b) Only new varieties with confirmation of DUS testing
(c) Only Public sector varieties after confirmation of DUS testing
(d) Extant and new varieties with confirmation of DUS testing

57. Which one of following is correct for the original variety from which the essentially Derived Variety is developed?
(a) May or may not be protected (b) Should be patented
(c) Should be protected (d) Should not be protected

58. If at the date of filing of the application for protection, the propagating material of such variety has not been sold with the consent of breeder or his successor for the purpose of exploitation of such variety earlier than one year in India before the date of filing such application. The Plant variety is considered as which one of following?
(a) Distinctive (b) Novel
(c) Stable (d) Uniform

59. Which one of following is considered as the variety submitted for protection?
(a) Candidate variety (b) Extant variety
(c) Farmers variety (d) Reference variety

60. Which one of following is the path for the stability of the new variety is tested/considered?
(a) Eberhart and Russell Model.
(b) Freeman and Perkins model.
(c) Perkins and Jinks model.
(d) Uniform and stable expression of the essential traits over the year and locations

61. Which one of following term is used for the deliberate plan of action to guide decisions and achieve rational outcome of seed?
(a) Act (b) Legislation
(c) Policy (d) Rule

62. Which one of following term is used for the law which has been enacted by government body to regulate, to authorize, to provide (funds), to sanction, to grant, to declare or to restrict?
(a) Act (b) Legislation
(c) Policy (d) Rule

63. Which one of following is considered for the principal changes include regulation and registration in new seed bill?
(a) All seeds to be sold, provisions for self-certification and accreditation of private seed testing laboratories, and regulation of transgenic seeds.
(b) Even transgenic seed can be sold after certification from public sector.
(c) For certification only private seed certification laboratory will be accredited.
(d) Seed can be sold only after certification from public sector, and no place for transgenic seed.

64. About the New Seed Bill 2004, which one of following is true?
(a) All seeds for sale must be registered.
(b) All varieties for sale must be registered.
(c) Only varieties notified by the government need to be registered.
(d) Registration is compulsory only for private seed agencies.

65. In New Seed Bill 2004, which one of following is considerable?
(a) No provision for transgenic varieties of seeds.
(b) Special provisions for registration of transgenic varieties of seeds.
(c) Transgenic varieties of seed will be register with non-transgenic without any discrimination
(d) Transgenic varieties of seed of Indian origin will be register with non-transgenic without any discrimination

66. New Seed Bill 2004 has which one of following In the event of underperformance of seeds?
(a) No specific provision for compensation.
(b) Provision for compensation to farmers under the Consumer Protection Act, 1986.
(c) Provision of cost of seed.
(d) Provision of seed replacement.

67. As per New Seed Bill 2004 any person who contravenes any provisions of the Act or imports, sells or stocks seeds deemed to be misbranded or not registered can be punishable by a fine of which one of following?
(a) Rs. 500-50000 (b) Rs 1000- 10,000
(c) Rs 5,000 to 25,000 (d) Rs. 10000/-

68. Seed Act Bill was introduced in the parliament on 29th Dec of which among the following year?
(a) 1965 (b) 1966 (c) 1967 (d) 1968

ANSWERSHEET

12.1 Seed Industry Development

1. c	2. a	3. b	4. d	5. d	6. c	7. a
8. c	9. d	10. b	11. d	12. d	13. b	14. d
15. c	16. b	17. c	18. d	19. a	20. a	21. b
22. d	23. a	24. b	25. d	26. c	27. b	28. d
29. c	30. d	31. d	32. b	33. d	34. b	35. a
36. c	37. c	38. c	39. c	40. b	41. d	42. b
43. d	44. c	45. b	46. a	47. c	48. c	49. b
50. b	51. d	52. d	53. d	54. d	55. b	56. d
57. a						

12.2 Seed Biology

1. d	2. c	3. b	4. a	5. c	6. a	7. b
8. c	9. b	10. d	11. c	12. b	13. d	14. a
15. d	16. d	17. a	18. d	19. b	20. d	21. a
22. b	23. b	24. b	25. b	26. c	27. c	28. b
29. a	30. d	31. b	32. d	33. d	34. a	35. d
36. a	37. b	38. c	39. a	40. b	41. c	42. c
43. b	44. d	45. d	46. d	47. c	48. c	49. a
50. a	51. d	52. d	53. c	54. a	55. a	56. a
57. c	58. b	59. c	60. d	61. d	62. c	63. b
64. a	65. b	66. d	67. d	68. a	69. b	70. d
71. a	72. a	73. a	74. c	75. a	76. d	77. b
78. d	79. b	80. d	81. c	82. d	83. b	84. c
85. a	86. d	87. d	88. a	89. a	90. d	91. b
92. a	93. b	94. b	95. c	96. d	97. a	98. d
99. d	100. b	101. b	102. a	103. a	104. b	105. b
106. a	107. d	108. c	109. c	110. d	111. c	112. a
113. a	114. b	115. c	116. d	117. a	118. b	119. a
120. c	121. c	122. b	123. c			

12.3 Seed Chemistry

1. b	2. d	3. b	4. b	5. a	6. d	7. d
8. c	9. c	10. d	11. a	12. b	13. a	14. d
15. d	16. a	17. d	18. c	19. c	20. b	21. a
22. a	23. b	24. d	25. d	26. c	27. a	28. a
29. d	30. a	31. a	32. b	33. a	34. b	35. c

12.4 Seed Health

1. d	2. a	3. a	4. c	5. c	6. b	7. b
8. a	9. b	10. c	11. c	12. b	13. d	14. a
15. b	16. b	17. d	18. d	19. a	20. c	21. a
22. b	23. a	24. d	25. b	26. d	27. b	28. c
29. a	30. b	31. d	32. a	33. d	34. b	35. b
36. a	37. a	38. c	39. c	40. b	41. c	42. d
43. c	44. a	45. c	46. a	47. a	48. b	49. b
50. a	51. b	52. c	53. b	54. c	55. d	56. d
57. d	58. d	59. d	60. b	61. d	62. d	63. c
64. d	65. d	66. b	67. b	68. b	69. d	70. d
71. b	72. a	73. b	74. d	75. b	76. d	77. a
78. b	79. b	80. b	81. b	82. a	83. b	84. a
85. d	86. b	87. b	88. a	89. c	90. d	

12.5 Seed Production Techniques

1. b	2. d	3. d	4. d	5. b	6. d	7. a
8. c	9. b	10. d	11. a	12. a	13. b	14. c
15. b	16. c	17. b	18. c	19. a	20. b	21. c
22. c	23. b	24. c	25. b	26. c	27. a	28. a
29. b	30. a	31. b	32. a	33. b	34. a	35. d
36. a	37. d	38. a	39. b	40. d	41. c	42. c
43. c	44. b	45. a	46. c	47. a	48. b	49. b
50. c	51. c	52. d	53. b	54. a	55. b	56. a
57. b	58. c	59. b	60. a	61. d	62. c	63. a
64. c	65. a	66. a	67. b	68. b	69. c	70. b

71. a	72. a	73. c	74. d	75. a	76. d	77. c
78. a	79. c	80. d	81. a	82. d	83. c	84. d
85. b	86. c	87. c	88. c	89. c	90. b	91. d
92. b	93. d	94. c	95. c	96. a	97. a	98. c
99. a	100. d	101. a	102. b	103. b	104. c	105. b
106. c	107. c	108. a	109. d	110. d	111. c	112. b
113. c	114. b	115. d	116. c	117. a	118. b	119. b
120. a	121. a	122. d	123. b			

12.6 Seed Processing

1. d	2. d	3. b	4. d	5. d	6. c	7. d
8. c	9. c	10. c	11. b	12. d	13. b	14. d
15. b	16. c	17. c	18. d	19. a	20. c	21. c
22. c	23. d	24. a	25. d	26. a	27. d	28. b
29. c	30. b	31. b	32. a	33. a	34. a	35. b
36. b	37. b	38. b	39. c	40. a	41. a	42. d
43. c	44. b	45. b	46. b	47. d	48. a	49. d
50. b	51. d	52. b	53. d	54. c	55. c	56. c
57. d	58. d	59. c	60. b	61. a	62. b	63. c
64. a	65. a	66. c	67. c	68. b	69. c	

12.7 Seed Quality Control

1. a	2. c	3. c	4. a	5. a	6. c	7. d
8. d	9. b	10. c	11. a	12. a	13. b	14. c
15. c	16. c	17. d	18. b	19. a	20. b	21. b
22. a	23. a	24. b	25. b	26. a	27. c	28. a
29. c	30. a	31. c	32. c	33. a	34. b	35. c
36. d	37. a	38. d	39. c	40. c	41. d	42. d
43. c	44. a	45. a	46. b	47. d	48. b	49. a
50. a	51. b	52. a	53. c	54. b	55. b	56. d
57. b	58. d	59. b	60. d	61. b	62. a	63. d
64. b	65. a	66. d	67. d	68. c	69. d	70. b
71. a	72. d	73. c	74. d	75. c	76. d	77. a
78. c	79. a	80. d	81. d	82. c	83. c	84. d
85. d	86. c	87. c	88. c	89. b	90. c	91. d
92. c	93. b	94. c	95. b	96. b	97. d	98. d
99. c	100. b	101. a	102. a	103. d		

12.8 Seed Storage

1. c	2. a	3. b	4. b	5. a	6. a	7. a
8. b	9. b	10. b	11. d	12. b	13. b	14. c
15. b	16. a	17. d	18. c	19. a	20. d	21. b
22. b	23. b	24. a	25. b	26. d	27. b	28. d
29. a	30. d	31. a	32. c	33. b	34. d	35. b
36. b	37. b	38. c	39. d	40. b	41. a	42. a
43. b						

12.9 Seed Certification

1.	d	2.	b	3.	c	4.	a	5.	c	6.	b	7.	a
8.	b	9.	b	10.	c	11.	b	12.	d	13.	c	14.	b
15.	a	16.	a	17.	c	18.	a	19.	d	20.	a	21.	c
22.	c	23.	c	24.	a	25.	a	26.	c	27.	c	28.	a
29.	c	30.	a	31.	d	32.	c	33.	c	34.	b	35.	a
36.	c	37.	a	38.	d	39.	d	40.	b	41.	a	42.	b
43.	c	44.	c	45.	c	46.	b	47.	b	48.	b	49.	b
50.	c	51.	b	52.	d	53.	d	54.	b	55.	b	56.	a
57.	c	58.	d	59.	d	60.	b	61.	d	62.	c	63.	a
64.	a	65.	b	66.	c	67.	a	68.	d	69.	d	70.	c
71.	d	72.	b	73.	d	74.	d	75.	b				

12.10 Seed Law Enforcement

1.	a	2.	b	3.	a	4.	c	5.	c	6.	a	7.	a
8.	a	9.	b	10.	c	11.	d	12.	c	13.	a	14.	a
15.	c	16.	d	17.	b	18.	b	19.	c	20.	a	21.	a
22.	a	23.	a	24.	a	25.	b	26.	b	27.	d	28.	b
29.	b	30.	c	31.	b	32.	d	33.	a	34.	b	35.	b
36.	c	37.	d	38.	d	39.	a	40.	a	41.	c	42.	b
43.	b	44.	c	45.	b	46.	d	47.	c	48.	b	49.	b
50.	c	51.	d	52.	d	53.	c	54.	d	55.	d	56.	d
57.	a	58.	b	59.	a	60.	d	61.	c	62.	b	63.	a
64.	a	65.	a	66.	b	67.	c	68.	b				

Glossary of Technical Terms

Abnormal Seedlings: Seedlings unable to develop into normal growth and healthy seedlings of plants in the field.

Abiotic Stress: Outside (non-living) factors which can cause harmful effects to plants, such as soil conditions, drought, and extreme temperatures.

Acclimatization: The physiological adaptation or adjustment of an individual plant or a population of plants under the changed in a complex of environmental factors such as a change in temperature, humidity, photoperiod, altitude or pH.

Achene: A small, hard, dry, one-chambered, one-seeded indehiscent fruit. The fruit wall is not joined with the seed coat. e.g. spinach and sunflower.

Allele or Allelomorph: An alternative form of a gene or genetic locus (a group of genes).

Allele: It is considered as several possible mutational forms of a gene at a given genetic locus. If a gene sits at the same place (locus) on a chromosome, the chromosome is to be said homologous. One diploid individual can have a maximum of two forms (alleles), but there may be several alternate alleles in the species.

Allopolypioidy: A polyploid in which all the sets of chromosomes are derived from the different species.

Aneuploid: An organism or a cell with a chromosome number that is not an exact multiple of the haploid chromosome complement.

Angiosperm: A group of flowering plants whose seeds is formed within a mature ovary (fruit). Today in the world, they are the dominant plants. All crop plants are of this plant class.

Anther: The saclike structure of the male part (stamen) of a flower in which the pollen is formed. Anthers normally have two lobes or cavities that dehisce at anthesis and allow the pollen to disperse.

Anthesis: The period of pollen distribution during which a flower is fully open and functional, may also refer to the onset of that period.

Apomixis: Apomixis is an asexual reproduction, genetically controlled reproductive process by which embryos and seeds develop in the ovule without female meiosis and egg cell fertilization. Apomixis was defined by Hans Winkler as replacement of the normal sexual reproduction by asexual reproduction, without fertilization. Apomixis produces seed progeny that are exact replicas of the mother plant. . It can also be used to describe the production of seeds without sex. There are different types of apomixes, they are; non-recurrent apomixis, recurrent apomixes (also called as gametophytic apomixes), sporophytic apomixes (also called as sporophytic budding, adventive embryony or nucellar embryony) and vegetative apomixes.

Asexual Reproduction: Is the process of producing new individuals by vegetative propagation, so it is also known as vegetative reproduction. It is a resulting in progeny, reproduction without involvement of meosis and syngamy which are clones or gametically identical to the parents and each other.

Autopolyploidy: An organism having more than two sets of chromosomes in body cells both sets derived from the same species.

Autosome: The chromosome other than sex chromosomes, as opposed to the sex chromosomes but never involve in sex determination because their kind and number is the same in males and females of a species.

Backcrossing: To cross of a offspring of genetically unlike parents (a hybrid organism) with one of its parents or with an individual genetically similar to one of its parents. This techniques is useful in plant breeding studies where the aim is not to create new varieties but for isolating (separating out) certain characteristics in a related group of plants to modify existing ones.

Biodiversity: Biodiversity is the degree of variation of life forms within a given species, ecosystem, biome, or planet. It is combination of the two words, "biological" and "diversity". Thus, in essence, biodiversity represents all life. India is one of the mega biodiversity centres in the world and has two of the world's 18 'biodiversity hotspots' located in the Western Ghats and in the Eastern Himalayas. The forest cover in these areas is very dense and diverse and of pristine beauty, and incredible biodiversity.

Clelstogamy: Cleistogamy or automatic self-pollination describes the trait of certain plants to propagate by using non-opening, self-pollinating flowers.

CMS (cytoplasmic male sterile) line: A line whose anthers produce no pollen or only abortive pollens. A genetic factor which induces sterility is present in its cytoplasm. No seed is set on this line by selfing. But its pistil is normal and can produce seeds when pollinated by a restorer or maintainer line.

Cross pollination: Also called allogamy. It occurs only when pollen grains are transferred from the anther of a flower to the stigma of a flower of a different plant of the same species.

Cross-fertilization Cross-fertilization is when the pollen travels to a different plant, increasing genetic diversity in the next generation.

Crossing Over: A cellular process of genetic recombination during meiosis where genes are exchanged of (usually) equivalent segments between homologous chromosomes.

Cultivar: A variety of plant or grouping of plants selected for desirable characteristics that can be produced and maintained through cultivation and propagation.

Cytoplasmic Inhertiance: Also called as extranuclear inheritance. The transmission of genes dependent upon cytoplasmic structures and not nuclear genes, occur outside the nucleus. It is generally found in cytoplasmic organelles (semi-autonomous) such as chloroplasts and mitochondria in most of the eukaryotes organisms that contain DNA.

Damping-off A single, collective term used to describe underground, soil line, or crown rots of seedlings due to unknown causes. Many soil borne diseases and fungi can cause these symptoms of sudden plant death.

Damping-off disease: A disease caused by one or a combination of two or more species of water molds. Evidence of damping-off disease is seen as cottony growths; the hyphae of the parasitic fungi are seen on the seed surface. The hyphae usually grow from a slit in the seedcoat that has been opened or broken during germination or threshing.

Deficiency: Any inadequacy or shortage of substances essential to growth and development of plants. Lacking in some quality, faculty, or characteristic necessary for completeness.

Dominance: Inter-allelic/intragenic interaction with complete suppression of one allele by another.

Ecotype: A subspecies or race that has developed distinct physiological characteristics like herbicide resistance, to have adaptation(s) in response to a specific environment which presumed even if individuals are moved to a different environment.

Emasculation: The systemic removal of the entire male sex organ (like anthers or kitting of pollen grains of a flower) without affecting the female reproductive organs in hermaphrodite flower. in rice, the removal of the stamen of the floret.

Endosmosis: It was derived from the now obsolete Latin word, endosmose. The process where movement of water of external environment into a cell inward from a state of lower to higher concentration through a semi permeable membrane as a result of greater internal osmotic pressure.

Epistasis: The phenomenon in which Interaction between genes, one gene suppresses to the action of another gene located at a different locus. The term 'epistatic' was first used in 1909 by Bateson to describe a masking effect whereby a variant or allele at one locus prevents the variant at another locus from manifesting its effect.

Eugenics: The applications of principles, theories and practices of genetics for the improvement of the genetic quality in human welfare. Further divide into two types, they are; 1) *positive eugenics*: refers to the improvement of genetic traits through the promotion of higher reproduction with desired traits and 2) *negative eugenics*: refers to the improvement of genetic traits through the promotion of reduced reproduction with less desired or undesired traits.

Evolution: It is a process of change in the inherited characteristics of biological (organisms) populations over successive generations.

Gamete: A mature reproductive male or female germ cell (sperm or egg) which is capable of uniting with another gamete to produce a zygote or the one-celled stage of an-embryo.

Gene: A functional or physiolological hereditary unit, which occupies a fixed location on the chromosomes within the nucleus of each cell, which determines to the heritable characteristics of the cell. It can be seen as a small bit of DNA segment that performs a specific function and is a unit of inheritance.

Genetic Drift: This is the random change in changes in the allele frequencies of a population when the size of sample chosen for rejuvenation is small. Genetic drift leads to a loss of certain genotypes in the population. It is important to note that differences in phenotype or genotype due to genetic drift do not reflect adaptive differences.

Genetic Engineering: Technologies used to isolate genes from an organism, manipulate them in the laboratory, and insert them stably into another organism. It refers to all of the techniques used to manipulate DNA, including gene isolation and cloning, gene splicing, plasmid construction, and transformation.

Genetic Erosion The loss of genetic variation, associated with a narrowing of the genetic base or a loss of heterozygosity. Can be used within the context of agricultural varieties where a narrow genetic base results following breeding or selection of favorable characteristics and the loss of other genes.

Genetic Purity: Trueness to type; seeds or plants genetic potential. The likelihood that the genetic material (improved or primitive) is capable of producing many new genotypic combinations from stored variousness by hybridization and subsequent segregation, recombination, and selection under varying pressures conforming to the characteristics of the variety, line, or hybrids.

Genome: A complete single set of the genetic material of a cell or organism; the complete set of genes in a gamete, the single DNA/RNA molecule of bacteria, phages, and most animal and plant viruses. In plants, composed of the nuclear genome, the mitochondrial genome, and the chloroplast genome.

Germination: The resumption of active growth of the embryo in the seed and the process by which a seed sprouts to development into a seedling when a favorable environment exists, culminating in the development of a plant from the seed.

Gynandromorph: An individual a part of whose body exhibits male sex characters and a part of female characters.

Heterosis / Hybrid Vigour: A method of breeding to develop F1 hybrids which exploit the phenomenon of hybrid vigor to increase yield potential and yield stability.

Homozygote: A zygote derived from the union of genetically identical gametes. An individual with both alleles alike for any given gene; these are therefore pure for any given trait.

Hybrid Variety: A variety developed from crossing between two purelines, inbreds, open pollinted varieties or other population which are genetically dissimilar, used for commercial planting to gain higher yields.

Hybrid: A plant obtained by crossing or mating between genetically dissimilar parents.

Hybridization: A method of breeding to produce new crop varieties in which two or more plants are crossed to generate new variability and to produce desired recombinants.

Improved Seed: A seed has been released as popular variety produced by scientific method. Nucleus seed, breeder seed, foundation seed, registered seed and certified seed are five types of improved seed.

In Vitro: Outside the living cell. in a vessel or test tube or other artificial medium.

In Vivo: Within the living organism.

Inbred Line: A nearly homozygous breeding line (pure line) originating by continued self-fertilization.

Inbreeding Depression: Continuous breeding of genetically related individuals resulting in decrease in reproductive capacity.

Inbreeding: Closely related individuals occurring naturally (as in a closed population), or as a deliberately chosen system of breeding and serving especially to preserve and fix desirable characters of and to eliminate unfavorable characters from a suitably selected stock but tending to effect an unwanted decline (as in size, vigor, or fertility) through the fixation of undesirable and often recessive characters when the initial stock is any way defective.

Indehiscent: Does not open at maturity to discharge seeds, not splitting open when ripe.

Independent Assortment: Random assortment in which segregation for one type has no effect on any other alleles. Meiotic behavior of genes located on non-homologous chromosomes.

Introgression: Incorporation of genes from one species or subspecies into another related species or subspecies by hybridization and backcrossing.

Inversion: A rearrangement of a chromosome segment so that reverse the order of genes on it. e.g. warm air near the ground, cool air above.

Irradiation Breeding: Use of irradiation for exposure of plants or plant parts to increase mutation rates with an objective of obtaining mutual plants for development of improved variety.

Karyotype: In an organism, the characteristics of entire set of chromosomes by which to the chromosomes number, size, shape and other features are described.

Lethal Gene: A gene or allele which causes sufficiently drastic to death the bearer of certain genotypes.

Open-pollinated Seed: A seed allows growers to harvest and save seed for the following year, which is produced progeny just like the parent plants.

Pedigree Record: A record register in which the relationship between parents and offspring are maintained by plant breeders.

Perfect Flower: Is also known as bisexual or hermaphrodite flower, which possesses both androecium and gynoecium.

Plant Breeder's Rights: A rights legally provided to the owner (plant breeder) of a variety for regulating the production and marketing of their variety.

Pollen: A fine mature reproductive bodies containing the microgametophytes (male gametes sperm cells) of seed plants, which are borne in the anthers of flowers.

Polycross: A cross used in selective plant breeding in which the female parent is known but the male may belong to any of several available strains, involves finding clones of strains that, upon crossbreeding with other clones or strains of the same species, yield the most productive plants. This method is used for developing synthetic varieties, successfully cloning (perennial herbs and annuals and biennials that propagate vegetatively) and also production of new lines of some economic plants (e.g. lucerne *Medicago sativa*)

Polymorphism: The occurrence of two or more phenotypes are clearly different in the same breeding population of a species, is maintained in a random mating population, due to the effect of environment.

Polyploid: The plant has with more than two completes sets (genornes) of chromosomes.

Prepotency: The capacity of a parent, variety, or strain to transmit its individual characters on its progeny because of possesses more homozygous, epistatic, or dominant genes.

Qualitative Character: A discrete heritable character or trait governed by polygenes, has discontinuous variation and transmitted well-defined limits and is in a simple alternate manner. e.g. number of seeds, height, total yield etc.

Quantitative Character: A heritable character or trait governed by one or few oligogenes, has continuous variation and readily affected by the environment. e.g. seed shape, seed color, seed form etc.

Recessive: the genetics relating to or designating an allele which cannot express a characteristic effect when present with a dominant allele, but can express only when the determining allele is present in the homozygous condition.

Reciprocal Crosses: A pair of crosses of between two parents where the phenotype of each sex reversed as compared with the original cross, to test the role of parental sex on inheritance pattern. e.g. Male Phenotype A x Female Phenotype ***B*** **and** Male Phenotype ***B*** x Female Phenotype A.

Recurrent Selection A method of breeding designed to reselection generation after generation of favorable genes for yield or other characteristics of selected plant to produce the population for the next cycle of selection. Hayes and Garber in 1919 and East and Jones in 1920 were initiated to the idea of recurrent selection. But the term was first coined by Hull in 1945. There are four types of recurrent selection, they are; 1) Simple recurrent selection (SRS), 2) Recurrent selection for general combining ability (RSGCA), 3) Recurrent selection for specific combining ability (RSSCA) and 4) Reciprocal recurrent selection (RRS).

Resistance: To suppress or retard the activities of a specified pest or pathogen by an ability of plants. In other hand, withstand a specific environmental or chemical stress by an ability of plants.

Rhizome: Modified underground, horizontal stem of a plant that is usually produce shoots and roots.

Roguing: The process of removal of phenotypically different type (off types) plants to avoid contamination from the field of an improved variety.

Scarify (scarification): The manual scratching or nicking of the seed coat using abrasion, thermal stress, or chemicals in order to promote/encourage fast germination.

Seed Dormancy: A period in which seeds are prevented or failed from completion of germination and survival of the seedlings under favorable environmental conditions such as complex combination of water, light, temperature, gasses, mechanical restrictions, seed coats, and hormone structures. There are different type of dormancy, they are; 1) Exogenous dormancy (e.g. physical dormancy, mechanical dormancy, chemical dormancy), 2) Endogenous dormancy (e.g. physiological dormancy, morphological dormancy, combined dormancy), 3) Combinational dormancy, 4) Secondary dormancy.

Seed: The product of fertilized ovule which consists of embryo, seed coat and cotyledon (s), usually with some stored food. It is used for commercial multiplication of a crop.

Segregation: During meiosis, the separation of two allelic genes into different meiospores or gametes.

Selection: A process by which the individual plant or the groups of plants are sorted out from mixed population.

Self-fertilization: Fusion of male and female gametes (sex cells) produced by a plant fertilizes that same plant or within a clone, maintaining or decreasing genetic diversity in the offspring.

Self-incompatibility: Is a widespread mechanism in flowering plants that prevents inbreeding (self-fertilization) and promotes outcrossing. The self-incompatibility response is genetically controlled by one or more multi-allelic loci, and relies on a series of complex cellular interactions between the self-incompatible pollen and pistil.

Selfing: Transfer of pollen from the male (anthers) to the female (stigma) of the same flower, also known as self-pollination. It is a mating with itself and important to know that does not always preclude cross pollination.

Self-pollination is the transfer of pollen grains from the anther to the stigma of the same flower

Self-pollinaton or Autogamy: When the pollen gains from a male (anthers) falls on receptive female (stigma) of the same flower.

Sex Chromosome: A chromosome, either of a pair of chromosomes that determine whether an individual is male or female. The sex chromosomes of human beings and other mammals are designated by scientists as X and Y, which differs in number or kind between the male and female individuals of the same species and are involved in sex determination (Syn., allosome). In humans the sex chromosomes comprise one pair of the total of 23 pairs of chromosomes while in the other, 22 pairs of chromosomes are called autosomes.

Sex-influenced Genes: Genes which are expressed more frequently in one sex than in other. For example, in humans pattern baldness is dominant in males and recessive in females.

Sex-linked Gene: It is a association of a gene or trait with sex, first discovered in Drosophila. Genes for such traits are located in sex chromosome. In XY (species gene are located only on the X chromsome.

Sucker: Is an unwanted shoots from the stem or roots which draw nutrients and give sap to the plants fruiting and flowering ability.

Susceptibility: Is an inability of plants which restricts to the activities of a specified pests or pathogens. It is also, an inability to withstand a specific environmental or chemical stress.

Sustainable Agriculture: To create integrated, humane, environmentally and economically sustainable production systems, which maximize reliance on farm-derived renewable resources and the management of ecological and biological processes and interactions, so as to provide acceptable levels of crop, livestock and human nutrition, protection from pests and disease, and an appropriate return to the human and other resources.

Test Cross: A cross between an individual exhibiting the dominant phenotype of a trait and an individual that is homozygous recessive for that trait in order to determine the genotype of the dominant individual.

Tetrasomic: A cell or an organism in a form of aneuploidy with the presence of four copies, instead of the normal two, of a particular chromosome.

Tetrazolium Test: A test for seed viability, have been developed to furnish quick estimates of seed's germinability. The test is very useful in processing, handling, storing and marketing. The large quantities of seed in a short time, testing dormant seed lots, vigour rating of the seed lots, supplernenting germination test results and diagnosing the cause of seed deterioration.

Tolerance: The capacity of plants to endure a specified pest, pathogen, environmental pressure or chemical stress. A susceptible variety sustains more damage than a tolerant variety when grown under the same conditions.

Totipotency: An ability of a single cell (such as an egg, protoplast) to regenerate into a new organism or part.

Trait: An organism with morphological, anatomical or physiological feature, yields by the actions of both genotype and environment.

Transgenic Plants: A crop plant in which a gene or genes from one organism have been artificially inserted into instead another unrelated plant (completely different species) through genetic engineering e.g. tobacco, tomato, brinjal etc.

Transgressive Variation: Appearance in offspring with more extreme expression of a phenotypes that exceed the parental extremes parents in the segragating generation.

Trisomic: A type of polysomy resulting of an error during cell division in which there are an extra copy of a particular chromosome of a set (2n + 1), instead of the normal two. They give a total of 47 chromosomes rather than the usual 46. They are an abnormal number of chromosomes, also known as a type of aneuploidy.

Variation Occurrence of differences among various species.

Variegated: Having irregular patches or streaks or spots which exhibiting in different colors.

Variety: A genotype or population of plants released for commercial cultivation, having many characteristics in common, a variety may be a pure line, a mixture of pure lines, a Mendelian population, or a clone. It is released either by State Variety Release Committee or Central Variety Release Committee.

Viability: An ability of a seed to capable of living, developing, or germinating when provide it to all environmental condition. It is usually expressed by percentage of viable seeds in a seed lot.

Wide Crossing: The process of undertaking a cross between two species of the same genus or of different genera, where one parent is from outside the immediate gene pool of the other, i.e. Landrace or primitive line crossed with a modern cultivar.

Xenia: Effect of the genotype of pollen grain on the phenotype of seed tissues.

Zygote: Beginning stage of an embryo which is an initial cell formed by the fusion of two gametes cells before mitosis or meiosis, joined by means of sexual reproduction. They are diploid number of chromosomes.

Suggested Books for Further Study

A text book on molecular genetics. Bharadwaj DN. 2008. Kalyani Publishers.

Advances in Plant Breeding. Mandal AK, Ganguli PK & Banerji SP. 1995. Vol.I, II.CBS.

An Integrated System of Classification of Flowering Plants. Cronquist AJ. 1981. Columbia Univ. Press.

An Introduction to Genetic Analysis. Griffiths AJF. 2000. WH Freeman.

An Introduction to Recombinant DNA Technology - Basic Experiments in Gene Manipulation. 2nd Ed. Hackett PB, Fuchs JA & Messing JW. 1988. Benjamin Publ. Co.

An Introduction to Seed Technology. Thomson JR. 1979. Leonard Hill, London.

Approaches for Incorporating Drought Salinity Resistance in Crop Plants. Chopra VL & Paroda RS. 1986. Oxford and IBH, Publishing Co., New Delhi.

Biotechnology, Expanding Horizons. Singh BD. 2005. Kalyani.

Breeding - Biotechnological and Conventional Approaches. Chahal GS & Ghosal SS. 2002.

Breeding Asian Field Crops. Poehlman JM & Borthakur D (1995) Oxford and IBH Publishing Co., New Delhi.

Breeding Field Crops. Chopra VL. 2001. Oxford & IBH.

Breeding of field crops. Bharadwaj DN. 2012. Agrobios India

Breeding of Horticultural Crops – Principles and Practices. Kumar N. 2006. New India Publishing Agency, New Delhi.

Cell and Molecular Biology: Concepts and Experiments. Karp G. 2004. John Wiley. Chapman & Hall.

Chromatin Structure and Gene Expression. Elgin SCR. 1995. IRL Press.

Chromosome Techniques: Theory and Practice. Sharma AK & Sharma A. 1988. Butterworth.

Crop Breeding and Genetics. Ram HH & Singh HG. 1993. Kalyani.

Cytogenetics of Aneuploids. Khush GS. 1973. Academic Press.

Cytogenetics. Gupta PK. 2000. Rastogi Publ.

Cytology and Cytogenetics. Swanson CP. 1960. Macmillan & Co.

Diversity and Classification of Flowering Plants. Takhrajan A. 1997. Columbia Univ. Press.

Drying of Farms Crops. Hall CW. 1966. Lyall Book Depot.

Elements of Biotechnology. Gupta PK. 1997. Rastogi Publ.

Essential Cell Biology. Bruce A. 2004. Garland.

Essential Genetics. Russell PJ. 1996. Blackwell Scientific Publ.

Essentials of Plant Breeding. Phundan Singh. 2006. Kalyani Publishers, New Delhi.

Fundamentals of Plant Breeding and Hybrid Seed Production. Agarwal RL. 1996. Oxford & IBH.

Genes and Genomes. Singer M & Berg P. 1991. John Wiley & Sons.

Genes IX. Lewin B. 2008. John Wiley & Sons.

Genetic Engineering and Biotechnology: Concepts, Methods and Applications. Chopra VL & Nasim A. 1990. Oxford & IBH.

Genetics and Molecular Biology. Schleif R.1986. Addison-Wesley Publ. Co.

Genetics. 4th Ed. Snustad DP & Simmons MJ. 2006. John Wiley & Sons.

Genetics. Singh BD. 2003. Kalyani.

Genetics. Strickberger MW. 2005. 3rd Ed. Prentice Hall.

Genomes. Brown TA. 2002. Wiley-LISS.

Handbook of Seed Testing. Agrawal PK. 1993. Ministry of Agriculture, GOI, New Delhi.

Heterosis and Hybrid Seed Production in Agricultural Crops. Basra AS. 2000. Food Product Press.

Hybrid Seed Production in Field Crops. Singhal NC. 2003. Kalyani Publishers, New Delhi.

Indian Minimum Seed Certification Standards. Tunwar NS & Singh SN. 1988. CSCB, Ministry of Agriculture, New Delhi.

Insect Pests of Stored Grain and Grain Products. Cottong RT. 1963. Burgess Publ. Co., Minneopolis, Minn., USA.

Introduction to Plant Breeding. Briggs FN & Knowles PF. 1967. Reinhold.

Legislation on Seeds. Anonymous. 1992. NSC Ltd., Department of Agriculture and Cooperation, Ministry of Agriculture, New Delhi.

Lehninger's Principles of Biochemistry. Nelson DL & Cox MM. 2005. WH Freeman & Co.

Methods of Plant Breeding. Hayes HK, Immer FR and Smith DC. 1955. McGraw-Hill.

Molecular Biology of the Gene. Watson JD. 2004. Pearson Edu.

Molecular Cell Biology. Lodish H, Berk A & Zipursky SL. 2004. 5th Ed.WH Freeman.

Objective Genetics and Plant Breeding. Singh P. 2002. Kalyani.

Physiology and Biochemistry of Seed Dormancy and Germination. Khan AA. 1977. North Holland Co.

Physiology and Biochemistry of Seeds in Relation to Germination. Bewley JD & Black M. 1982. Vols. I, II. Springer Verlag.

Plant Biotechnology. Mantel SH & Smith H. 1983. Cambridge University Press.

Plant Breeding for Stress Environments. Blum A. 1988. CRC Press.

Plant Breeding. Chopra VL. 2004. Oxford & IBH.

Plant Breeding. Singh BD. 2006. Kalyani Publishers, New Delhi.

Plant Breeding: Principles and Methods. Singh BD. 2006. Kalyani Publishers, New Delhi.

Plant Genetic Resources Conservation and Management Concepts and Approaches. Paroda RS & Arora RK. 1991. IPGRI Regional office for South and South Asia, New Delhi.

Plant Systematics: An Integrated Approach. Gurcharan Singh. 2004. Science Publ.

Plant Taxonomy and Biosystematics. 2nd Ed. Stace CA. Cambridge Univ. Press.

Practical Manual of Floral Biology of Crop Plants. Chhabra AK. 2006. Department of Plant Breeding.CCS HAU Hisar.

Practical Manual on Basic and Applied Genetics. Tamarin RH. 1999.

Practical Plant Breeding. Gupta SK. 2005. Agribios.

Principles and Practice of Plant Breeding. Sharma JR. 1994. Tata McGraw Hill, Publishing Company Ltd., New Delhi.

Principles of Crop Improvement. Simmonds NW. 1990. English Language Book Society.

Principles of Genetics. Gardner EJ & Snustad DP. 1991. John Wiley & Sons.

Principles of Genetics. Tamarin RH. 1999. Wm C Brown Publ.

Principles of Plant Breeding. Allard RW. 1960. John Wiley and Sons, New York.

Principles of Seed Certification and Testing. Nema NP. 1986. Allied Publs.

Principles of Seed Pathology. Agarwal VK & Sinclair JB. 1997. Boca Raton.

Principles of Seed Processing. Virdi SS & Gregg BG (1970) National Seed Corp., New Delhi.

Principles of Seed Science and Technology. Copeland LO & McDonald MB. 2001. 4th Ed. Chapman & Hall.

Principles of Seed Technology. Agarwal PK, 1994. ICAR, New Delhi.

Seed Hand Book: Biology, Production, Processing and Storage. Desai BB, Katecha PM & Salunke DK. 1997. Marcel Dekker.

Seed Identification Manual. Martin C & Barkley D. 1961. Oxford & IBH.

Seed Production of Agricultural Crops. Kelly AF. 1988. Longman.

Seed Production of Commercial Vegetables. Singh SP. 2001. Agrotech.

Seed Production: Principles and Practices. McDonald MB Jr & Copeland LO. 1997. Chapman & Hall.

Seed Quality: Basic Mechanisms and Agricultural Implications. Basra AS. 1995. Food Products Press.

Seed Technology. 2nd Ed. Agarwal RL. 1997. Oxford & IBH.

Seed Technology. Agarwal RL. 1996. Oxford and IBH Publication Co., New Delhi.

Taxonomy of Vascular Plants. Lawrence GMH. (Ed.). 1951. London.

Techniques in Seed Science and Technology. Agarwal PK & Dadlani M. 1986. South Asian Publishers, New Delhi.

The Handling of Chromosomes. Darlington CD & La Cour LF. 1969. Georger Allen & Unwin Ltd.

The Science of Genetics. Hexter W & Yost HT. 1976. Prentice Hall.

The World of Cell. Carroll M. 1989.

Vegetable Breeding. Kalloo G. 1994. Panima Educational Book Agency, New Delhi.